Encyclopaedia of

Genes, Genomes, Genetics and Chromosomes

Encyclopaedia of

Genes, Genomes, Genetics and Chromosomes

Volume 3

Editor:
Dr. R.N. Shukla

AGROTECH PRESS
JAIPUR ▪ NEW DELHI

AGROTECH PRESS

Regd. Off. "Bony Residency", Gate No. 2
Opp. Tilak Public School, Vishweswariya Nagar,
Gopalpura Road, Jaipur - 302018, India
Ph: 0141 - 2761280, Telefax: 2761381
E-mail : agrotechpress@gmail.com

Delhi Office: 102, 1st Floor, Satyam House,
4327/3, Ansari Road, Darya Ganj,
New Delhi - 110002, India
Ph: 011-45652440
visit us: www.agrotechpress.com

Distributed By
KEYA INTERNATIONAL PRESS LLP
E-mail: keyainternationalpressllp@gmail.com

Published in India, 2024

ISBN : 9789384568238 (5 Vols. Set) _3

Typeset by Thomson Press

Printed and bound in India by Thomson Press Ltd.,
New Delhi

Contents

Preface

Genetic engineering, also called genetic modification, is the direct manipulation of an organism's genome using biotechnology. New DNA may be inserted in the host genome by first isolating and copying the genetic material of interest using molecular cloning methods to generate a DNA sequence, or by synthesizing the DNA, and then inserting this construct into the host organism. Genes may be removed, or "knocked out", using a nuclease. Gene targeting is a different technique that uses homologous recombination to change an endogenous gene, and can be used to delete a gene, remove exons, add a gene, or introduce point mutations. An organism that is generated through genetic engineering is considered to be a genetically modified organism. The first GMOs were bacteria in 1973; GM mice were generated in 1974. Insulin-producing bacteria were commercialized in 1982 and genetically modified food has been sold since 1994.

Gene technology is the term given to a range of activities concerned with understanding the expression of genes, taking advantage of natural genetic variation, modifying genes and transferring genes to new hosts. Gene technology sits within the broader area of biotechnology – the use of living things to make or change products. Humans have been using biotechnology for centuries in activities ranging from plant and animal breeding through to brewing and baking. As our understanding of how living things function, grow and reproduce increases, modern biotechnology creates new opportunities for food and fibre production. Modern biotechnology includes the discovery of genes, understanding how genes function and interact (functional genomics), discovery of natural DNA markers to select more efficient plants and animals and genetic modification or genetic engineering. Genes have three basic parts – an 'on' switch, an 'off' switch and the region that codes for a protein. For a gene to work, it must have all three basic parts, but they do not have to be from the same organism. This is because the DNA of all living things is made up of the same four molecular components called nucleotides. Each of the four nucleotides has a different base, called adenine, guanine, thymine and cytosine.

This means the gene for a particular characteristic that has been identified in one organism can be permanently introduced into another organism after being modified so it will be recognised by the new organism, which then adopts this new characteristic. For example, you could take the protein-producing part of a gene in corn that produces vitamin A and combine this with the on and off gene switches from a seed gene from rice. The result would be a gene that produces more vitamin A in rice seed. Gene technology is a tool that offers potentially enormous benefits, but since it is still a relatively new technology, it is important that use of genetically modified organisms and products from GMOs are carefully regulated. Gene technology is helping scientists to develop more effective therapies for diseases like cancer, diabetes, hepatitis C and influenza. For instance, micro-organisms such as yeast and bacteria have been modified to produce vaccines for hepatitis B and insulin for diabetics. Researchers in Australia have also developed a malaria vaccine by genetically modifying the malaria parasite *Plasmodium falciparum*, which is now undergoing trials. Scientists also use gene technology to locate and study the genes that cause genetic diseases, or that make some people prone to heart disease, Alzheimer's disease, motor neurone disease, some cancers, rheumatoid arthritis and lupus. Humans have altered the genomes of species for thousands of years through artificial selection and more recently mutagenesis. Genetic engineering as the direct manipulation of DNA by humans outside breeding and mutations has only existed since the 1970s. The term "genetic engineering" was first coined by Jack Williamson in his science fiction novel *Dragon's Island*, published in 1951, one year before DNA's role in heredity was confirmed by Alfred Hershey and Martha Chase, and two years before James Watson and Francis Crick showed that the DNA molecule has a double-helix structure.

This book outlines the growth, development and prominence of this science. Highlighting the principles involved in the subject, it gives a comprehensive analysis of the numerous issues, the various techniques, and the current trends and developments in the field. The subject matter has been presented in such a way that it is easily accessible to the students, teachers and researchers.

—*Editor*

1

DNA Quantification and Quality Assessment

DNA quantification and quality assessment is done by using UV-VISspectrophotometry.Normally quality check is performed through the A260/ A280 ratio that is 1.8 value shows the highest purity, if more than 1.8 shows the presence of RNA contamination and less than that shows the protein contamination (Sambrook, *et. al,* 1989).

AGE is conducted of isolated DNA samples for visualization of bands. For perfect extraction, there will be clear single bands otherwise smear like image will be seen. That is in case of egraded DNA. For PCR based methods the DNA samples are diluted for PCR equirements. The optimum concentration is usually 40- 60ng/µL.

Polymerase Chain Reaction (PCR)

The DNA amplification by thermal cycling called Polymerase Chain Reation is in vitro method that can be used to amplify a specific DNA segment from small amounts of DNA template or duplex into millions of copies. It is invented by Kary Mullis *et.al,* (1985).

Heatdenaturation

This temperature denatures the double stranded DNA into two individual stands. Denaturation temperature is 95^0 C for 30 seconds or is 97^0 C for 15 seconds, however higher temperature may be appropriate, especially for G+C rich nucleotides.

Annealing

During this time one primer binds with the 5 prime end of one DNA strand and the other primer binds with 3 prime end of its complementary strand. Annealing is hybridization of primers to single stranded DNA and the length of time required for primer annealing depends on the basic composition, length and concentration of primers.

Primer Extension

This temperature is varies for Taq DNA polymerase which adds complementary nucleotides one by one to the 3' OH group of the primer. Estimates for the rate of nucleotide incorporation at 72^{0}C vary from 35-100 nucleotides per second depending upon the buffer, pH salt concentration and nature of DNA template.

Number of PCR Cycles

The optimum number of cycles depends mainly upon the starting concentration of target DNA, when other parameters are optimized. A common mistake is to execute too many cycles. The increase in the number of cycles will increase the amount and complexity of non-specific background products.

Restriction Fragment Length Polymorphisms (RFLPs)

RFLP Definition

The variation(s) in the length of DNA fragments produced by a specific restriction endonuclease from genomic DNAs of two or more individuals of a species (Kahl, 2001).

Restriction fragment length polymorphisms (RFLPs) was the first to develop (Soller and Beckmann, 1983) and are still widely used marker system especially in practical plant breeding. It was first developed in the 1980s for use in human genetic applications and was later applied to plants Most of the studies in systematic and evolution (Whitkus *et al.*, 1994) and molecular maps of different crop species were initially made through the use of RFLP markers. Restriction fragment length polymorphism (RFLP) technology. RFLPs are relatively small in size and are co-dominant in nature. Restriction fragments of different lengths are thus generated. All RFLP markers are analysed using a common technique. However, the analysis requires a relatively complex technique that is time consuming and expensive.

The hybridization results can be visualized by autoradiography (if the probes are radioactively labelled), or using chemiluminesence (if non-radioactive, enzyme-linked methods are used for probe labelling and detection). Any of the visualization techniques will give the same results. The visualization techniques used will depend on the laboratory conditions. A comprehensive study on these maps is documented by Phillips and Vasil (1994). Thus, RFLP markers have the potential to address with power, precision and accuracy, all issues of plant breeding in which markers can be applied. In addition to that, voluminous literature is available on utilization of RFLPs in other aspects especially in DNA fingerprinting (Tanksley, 1983), phylogenetic analysis (Zheng *et al.*, 1994), detection of genetic variability in rice, genetic dissection of maturity (Phillips *et al.,* 1992), identification of chromosomal regions responsible for regeneration of RFLPs, comparative mapping of chromosomes in wheat, rye and barley, analysis of wide compatibility in rice varieties (Zheng *et al.*, 1994), QTL mapping, gene tagging, and identification of ecotypes (Pupilli *et al.,* 2000).

To discover RFLPs, restriction enzymes (RE) are used to cut DNA at specific 4-6 bp recognition sites. Sample DNA is digested with one or more RE's and resulting fragments are separated according to molecular size using gel electrophoresis. After that Southern Blotting is performed where fragments are immobilized on a nylon membrane, followed by hybridization to a labelled probe (specific chemolu-minescence DNA probe) that identifies the locus under autoradiograph. Fragment sizes of digested DNA will differ depending on the presence or absence of the proper recognition sequence for the enzyme.

Differences in fragment length result from base substitutions, additions, deletions or sequence rearrangements within RE recognition sequences. Although two individuals of the same species have almost identical genomes, they will always differ at a few nucleotides. Some of these differences will produce new restriction sites (or remove them), and the banding pattern seen on a genomic Southern will thus be affected. For any given probe (or gene), it is often possible to test different restriction enzymes until one which gives a pattern difference between two individuals is found, that is, a RFLP. The less related the individuals, the more divergent their DNA sequences are and the more likely you are to find a RFLP.

However, the low level of polymorphism present in RFLPs restricted the use of this technique. In addition, RFLP requires large quantities of relatively pure DNA, and the frequent use of radioisotopes in detection step makes it technically demanding, laborious and costly to characterize large number of samples.

Utilization of RFLP markers requires a lot of skill in using short-lived radioactive material: the availability of which often becomes a limiting factor especially in the developing countries. Another limiting factor could be the large sample size to be analysed like that of F segregating population. In this case, utilization of RFLP markers may not be feasible, as it requires extraction of DNA in ultra pure form, restriction digestion, and lengthy and expensive process of isotopic labeling. Such difficulties have opened up a debate on the feasibility of routine application of RFLPs in large scale crop improvement programme. Another difficulty frequently encountered by RFLP users is the detection of low (1-3 loci) ornon-existent polymorphism in closely related species/varieties (Waugh and Powell, 1992). It was essentially this difficulty that led to the search for a new marker system.

Randomly Amplified Polymorphic DNAs (RAPDs)

Randomly-amplified polymorphic DNA markers (RAPD). In 1991 Welsh and McClelland developed a new PCR-based genetic assay namely randomly amplified polymorphic DNA (RAPD). This procedure detects nucleotide sequence polymorphisms in DNA by using a single primer of arbitrary nucleotide sequence.

In this reaction, a single species of primer anneals to the genomic DNA at two different sites on complementary strands of DNA template. If these priming sites are within an amplifiable range of each other, a discrete DNA product is formed through thermocycic amplification. On an average, each primer directs amplification of several discrete loci in the genome, making the assay useful for efficient screening of nucleotide sequence polymorphism between individuals. However, due to the stoichastic nature of DNA amplification with random sequence primers, it is important to optimize and maintain consistent reaction conditions for reproducible DNA amplification. RAPDs are DNA fragments amplified by the Polymerase Chain Reaction (PCR) using short (generally 10 bp) synthetic primers of random sequence.

Because of the short primers used, the reannealing temperature in the PCR must be low (35-40°C) for the primer to bind. However, due to low temperature, the binding is not very specific, which means that primers will bind also to sequences which are not completely complementary. These oligonucleotides serve as both forward and reverse primer and usually are able to amplify fragments from 3-10 genomic sites simultaneously. Amplified fragments (within the 0.5-5 kb range) are separated by gel-electrophoresis and polymorphisms are detected as the presence or absence of bands of particular size. These polymorphisms are considered to be primarily due to variation in the primer anealing sites.

They are dominant markers ad hence have limitations in their use as markers for mapping, which can be overcome to some extent by selecting those markers that are linked in coupling. RAPD assay has been used by several groups as efficient tools for identification of markers linked to agronomically important traits, which are introgressed during the development of near isogenic lins. The application of RAPDs and their related modified markers in variability analysis and individual-specific genotyping has largely been carried out, but is less poular due to problems such as por reproducibility faint or fuzzy products, and difficulty in scoring bands, which lead to inappropriate inferences.

In RAPD analysisthe total DNA from an organism is mixed with (generally) 10 bp single stranded DNA (primer) together with the four different doxynucleotides and a heat stable NA polymerase enzyme. The reaction mixture is placed on a thermocycler (PCR-machine) which can change the temperature of the reaction rapidly according to a redefined PCR programme.

Reannealing at low temperature makes the primer molecules attach to complementary site on the DNA. Elongation is the period when the polymerase enzyme elongates the attached primers from their 3' end. Denaturation separates the new formed trands. After a few PCR cycles DNA pieces with well defined length similar to the one between the two original primer sites dominate the mixure. This is because they amplify exponentially.

In most RAPD reactions the result will be amplification of 5 to 10 different DNA segments whose lengths depend on the primer recogition sites available in the genome. The DNA after amplification is visualized for bands using Agarose Gel Electrophoresis (AGE). The band scoring is done by comparing with the marker bands.

Inter Simple Sequence Repeats (ISSR)

ISSR amplification definition: variant of the polymerase chain reaction that uses simple sequence repeat primers (e.g. [AC] n) to amplify regions betwen their target sequences (Kahl, 2001).

Inter simple sequence repeat markers (ISSR). In this technique, reported by Zietkiewicz et al. primers based on microsatellites are utilized to amplify inter-SSR DNA sequences. Here, a single primer or combination of primers composed of a microsatellites sequence anhored with a single base at 3' region so that it can amplify regions betwee two microsatellites loci with oligonucleotides annealing to repeat sequence.

More complex banding patterns can be achieved using 5'-anchored primers that incorporate the SSR regions in their amplification products, and by combining 3'- and 5'- primers.ISSR circumvents the requirement for flanking sequence information, and thus has found wide applicability in a variety of plants.(Wolfe et al.1998} These are mostly dominant markers, though occasionally a few of them exhibit codominance.

An unlimited number of primers can be synthesized for various combinations of di-, tri-, tetra- and pentanucleotides [(4)3 = 64, (4)4 = 256] etc. with an anchor made up of a few bases and can be exploited for a broad range of applications in plant species. ISSR analysis is technically simpler than any other marker systems. The method provides highly reproducible results and generates higher levels of polymorphism compared to other markers in many systems. The advantage lies in long primer length and stringent annealing temperature. Primers used (from Yang *et al.,* 1996).

For separation and visualization of bands Agarose gel electrophoresis is conducted.

Simple Sequence Repeats (SSRs)

SSR (Microsatellites) Definition

Any one of a series of very short (2-10 bp), middle repetitive, tandemly arranged, highly variable (hyper variable) DNA sequences dispersed throughout fungal, plant, animal and human genomes (Kahl, 2001).

Simple sequence repeats (SSR) or microsatellites are a class of repetitive DNA elements. The di-, tri- or tetra-nucleotide repeats are aranged in tandem arrays consisting of 5 – 50 copies, such as (AT) 29, (CAC) 16 or (GACA) 32. SSRs are abundant in plants, occurring on average every 6-7 kb (Cardle *et al.*, 2000). These repeat motifs are flanked by conserved nucleotide sequences from which forward and reverse primers can be designed to PCR-amplify the DNA section containing the SSR.SSR alleles, amplified products of variable length, can be separated by gel electrophoresis and visualized by silver-staining, autoradiography (if primers are radioactively labelled)or via automation (if primers are fluorescently lbelled). SSR analysis is amenable to automation and multiplexing, and allows genotyping to be performed on large numbers of lines, and multiple loci to be analysed simultaneously. SSRs can be identified by searching among DNA databases (e.g. EMBL and Gene bank), or alternatively small insert (200-60bp) genomic DNA libraries can be produced and enriched for particular repeats (Powell*et al.*, 1996). From the sequence data, primer pairs (of about 20 bp each) can be designed (software programmes are available for this).

For separation and visualization of bands Polyacrylamide Gel Electrophoresis (PAGE) is conducted (6% acrylamide solution), stained and the observed the bands.

Amplified Length Polymorphisms (AFLPs)

AFLP definition: Any difference between corresponding DNA fragments from two organisms A and B that is detected by the amplified restriction length polymorphism technique (Kahl, 2001).

Table: *DNA Fingerprinting Techniques: A Comparative Analysis*

Marker	*Advantages*	*Disadvantages*
RFLP	Unlimited number of loci	Labour intensive
	Co-dominant	Fairly expensive
	Many detection systems	Large quantity of DNA needed
	Can be converted o SCARs	Often very low levels of polymorphism
	Robust in usage	Can be slow (often longexposure times)

	Good use of probes fromother species	Needs considerable degree ofskill
	Detects in related genomes	No sequence iformationrequired
RAPD	Results obtained quickly	Highly sensitive to laboratorychanges
	Fairly cheap	Low reproducibility within andbetween laboratories
	No sequence informationrequired	Cannot be used acrosspopulations nor across species
	Relatively small DNAquantities required	Often see multiple loci
	High genomic abundance	Dominant
	Good polymorphism	
	Can be automated	
ISSR	Highly polymorphic	Usually dominant
	Robust in usage	Species-specific
	Can be automated	
SSR	Fast	High developmental andstartup costs
	Highly polymorphic	Speces-specific
	Robust	Sometimes difficult interpretation because of stuttering
	Can be automated	Usually single loci even in polyploids
	Only very small DNA	<!—[if !support Misaligned Columns]—> <!—[endif]—>
AFLP	Small DNA quantitiesrequired	Evaluation of up to 100 loci
	No sequence informationrequired	Marker clustering
	Can be automated	Dominant
	Can be adapted for differentuses, e.g.	Technique is patented
cDNA-AFLP	Can be technically challenging	

This genetic marker system basically uses restriction fragment analyss, but a PCR approach is used to visualize fagments instead of hybridization with a probe. Because of the PCR step, this type of markers has characteristics quite differen from RFLP markers. Initially, DNA is extracted as for RFLP and then cu with two different restriction enzymes to produce well defined restriction fragments with sticky ends. Synthetic double stranded likers of approximately 18 to 20 bp with matching sticky nds are ligated on all the restriction fragments. Ligated fragments are subsequently amplified n PCR with 18 to 20 nucleotide length primers recognizing linkers in each end of the fragments.

A primer used for amplification in AFLP is normally labelled with P^{33} or some nonradioactive labeling system. The mixture of amplified fragments are separated according to size in a Polyacrylamide ge and visualized by means of autoradiography or other procedures for development of non radioactive systems. Alternatively, automated sequncing machines can be used toread the sequence of amplification products directly from the Polyacrylamide gel if primers have been labelled with special fluorescent dyes. Also, silver staining of the Polyacrylamide gel after electrophoresis can be used to stain all amplified DNA fragments in the gel. To avoid that all restriction fragments from the genome amplify in the same PCR, which will produce a smear because of too many DNA fragments, a two step amplification procedure is used. Firstly, part of the total number of restriction fragments are amplified in the "preamplification" with primers containing one extra "selective" nucleotide on their 3' end.

This selective nucleotide will allow amplification only of restriction fragments with a matching nucleotide next to the linker. With one selective nucleotide on both primers only 1/16th of all restriction fragments in the mixture will amplify during the preamplification. During the subsequent "selective" amplification, additionally, one or two selective nucleotides on each primer will further reduce the number of fragments amplified. A good AFLP amplification will show 90 to 100 different fragments in one analysis.

Applications

There are a variety of applications in DNA fingerprinting, such as to identify genetic diversity within breeding populations in plants and animals, to differentiate between plant species cultivars, as well as to identify plants containing a gene of interest., it is also used to detect genetically modified organisms in agriculture in the United States.

RFLP is most suited to studis at the intraspecific level or among closely related taxa. Presence nd absence of fragments resulting from changes in recognition sites are used for dentifying species or populations. In plant genetic analysis, RFLP markers were initially used for estimating genetic distance and fingerprinting in wheat. In plant and animal breeding, they can reveal which backcross progeny most resemble the desired parent. RFLPs are used t settle cases where biological parentage is at issue.

Many plant species are being analysed using RAPD method. RAPD markers were also used for characterisation, estimation of geetic relatedness and determination of genetic diversity of tea germplasm. These markers are able to distinguish between genotypes but limited to comparisons of populations from a few sources. Lastly RAPD analysis was used to assess geneic differences and relatedness among fig, *Ficus carica* L. clones.

For AFLP analysis, it is an fficient technology for assessing genetic diversity among cultivars such as wheat, *Triticum aestivum* cultivars. Cultivars were improved through crossing with elite breed however, crossing within such germplasm may lead to a reduction in the level of genetic variation among parental lines due to intense selection pressure. Thus DNA marker systems are used to reflect actual DNA differences among lines since selection pressure and genetic drift are accounted for. It is an efficient genetic marker system as the technique gives a much hiher degree of polymorphism than RAPD or RFLP. In addition, its prior sequence knowledge is not required as the fragments detected are selected by specific primer sets in the PCR reaction. It is reliable as stringent reaction conditions are applied for primer annealing.

Inter SSR (ISSR) fingerprinting was developed such that no sequence knowledge was required. Primers based on a repeat sequence, such as (CA) n, can be made with a degenerate 3'-anchor, such as (CA) 8RG or (AGC) 6TY. The resultant PCR reaction amplifies the sequence between two SSRs, yielding a multilocus marker system useful for fingerprinting, diversity analysis and genome mapping. Previous investigators have demonstrated that ISSR analysis usually detects a higher level of polymorphism than that detected with restriction fragment length polymorphism (RFLP) or random amplified polymorphic DNA (RAPD) analyses.

As with AFLPs, the great advantage of SSRs analysis is the large number of polymorphisms that the method reveals, with one locus in soybean (Glycine max) reported to have 26 alleles. Unlike AFLPs, microsatellites are co-dominant markers, thus heterozygotes can be readily identified. Microsatellite co-dominance will increase the efficiency and accuracy of population genetic measures based on these markers compared with other markers, such as AFLPs and RAPDs.

With a beginning as a mere forensic tool, the world of DNA fingerprinting has a gone a long way in revealing the genetic identity of living beings. With massive eolution in concepts and techniques it has given a lot to various fields cutting across the spectrum. In plants it has not only helped in identifying species but alsoin defining a new realm in plant genomics, plant breeding and in conserving the biodiversity. With world paving way for developments in biotechnology, DNA fingerprinting promises a very powerful tool in our future endeavors.

Intellectual Property Rights in Plant Varieties

Legal and Policy Concepts Relating to the Protection of Plant Varieties and Plant Breeders' Rights

Intellctual property rights (IPRs) are legal rights granted by governmental authorities to control certain products of human intellectual effort and ingenuity. (OECD 1996). An in-depth discussion of the philosophical and policy goals served by granting legal protection to these products is beyond the scope of this study. However, a basic familiarity with these goals is necessary to grasp how national and international intellectual property systems and institutions have evolved into theirpresent forms and to understand the constraints that those systems and institutions place o governments seeking to implement competing policy objectives in tension with IPRs. Two broad philosophical approaches underlie the decision to grant IPRs in the products of human intellectual effort and ingenuity. Elements of these two approaches can be found to different degrees in all national laws and international agreements relating to IPRs.

The first approach to IPR protection predominates in many civil law legal systems, including continental Europe.It takes the position that the products of the human mind are stamped with the personality of their creator, inventor or author, thus endowin him or her with a moral as well as an economic claim to exploit those products to the exclusion of third parties. Under this view, legal protection flows from a state's commitment to protect human rights, fact reflected in the wording of article 27 of the Universal Declaration of Human Rights, which guarantees to everyone "the right to the protection of the moral and material interests resulting from any scientific, literary or artistic production of which he is the author." Legal protection also rewards creators, inventors and authors for their intellectual efforts as well as their expenditures of time and money. From these underlying premises flows a desire to provide robust intellectual property protecton that includes, for example, a broad and expansive range of exclusive rights, long terms of protection, limits on mandatory licensing and narrow exceptions and limitations to exclusive rights.

The second approach to IPR protection takes as its starting premise an instrumental view of intellectual property. Legal protection for the products of human intellectual effort and ingenuity is granted not because of a moral commitment to compenating creators or innovators, but rather because the products they createenrich a society's culture and knowledge and thus increase its welfare. Perhaps the most well-known manifestation of this approach is found in the Intellectual Property Clause of the United States onstitution, which authorizes the United States Congress "[t]o promote the progress of science and useful arts, by securing for limited times to authors and inventors the exclusive right to their respective writings and discoveries."

This instrumentalist philosophy shapes the structure of many national intellectua property systems. The grant of IPRs in nations following an instrumentalist approach is intended to provide adequate incentives for creators, inventors and authors to invest the time, resources and intellectual capital needed to create intellectual property products.

In the absence of a grant of exclusive rights over those products, so-called "free riders" who have not made such investments could exploit inexpensive distribution and reproduction technologies and sell others' intellectual property products at a much lower cost. However, the ultimate goal of legal protection is not remunerative reward for creators but the enhancement of social welfare through access to the ideas and information contained in their products.

Instrumentalist intellectual property systems often tailor the scope of legal protection to achieve this goal, for example by placing certain limits on the scope of protection or recognizing situations in which consumers or second-generation creators may access and use intellectual property products forsocially valuable purposes.

Policy Objectives Favouring IPRs in New Plant Varieties

The policy goals of granting IPRs to plant varieties are grounded prinipally on an instrumentalist approach to IPRs. This is true both for patents and plant breeders' rights. Under this instrumentalist approach, new plant varieties are afforded legal protection to encourage commercial plant breeders to invest the resources, labour and time needed to improve existing plant varieties by ensuring that breeders receive adequate remuneration when they market the propagating material of those improved varieties. In the absence of a grant of exclusive rights to breeders, the dangers of free riding by third parties would be considerable. This is because the genetic material within plants that specifies their distinctive and commercially valuable features is naturally self-replicating, for example by reproduction of seeds or other propagating material. Self-replication maks innovations incorporating biological material particularly susceptible to exploitation by parties other than the innovator.

IPRs in plant varieties thus provide some assurance to breeders that they will be able to recoup the risks and costs of a value-added innovation that is based upon an underlying biological resource. Ultimately, however, the grant of exclusive rights to plant breeders is designed to benefit the society granting the rights. It provides an incentive for private research and development into new breeding techniques, thereby reducing the need for government funding to subsidize these activities. It encourages the development of new and beneficial plant varieties for use by farmers and consumers. And it furthers the society's development of agriculture, horticulture and forestry.

An international system of IPR protection for plant varieties expands these benefits by facilitating access to new varieties created in other states. Once breeders are assured that their rights will be protected in other states, breeders will be more willing to make their new varieties available in those states (assuming they have access to a distribution and marketing infrastructure). This benefits farmers, consumers and researchers in many more jurisdictions.

The Evolution and Structure of the International Intellectual Property System

The different policy objectives underlying the protection of IPRs have shaped the structure and evolution of the international intellectual property system. Most early domestic intellectual property laws provided no legal protection to intellectual property productscreated in other nations, thereby permitting tose products to be exploited by free riders operating outside the state in which the products were created. The unfairness of this result prompted governments in the late nineteenth century to consider an international approach to protect IPRs.

Limited treaty obligations : The drafters of the first multilateral intellectual property treaties quickly realized, however, that there was insufficient political support for reconciling many of the differences that existed among national IPR laws. For this reason, the drafters abandoned the idea of harmonizing diverse national laws to create a single, international IPR applicable in all signatory states. They fashioned instead a system that creates a limited set of treaty-based obligations that each member state of that system is then required to implement in its national IPR laws.

Implementation of treaty-based obligations in national IPR laws can occur in one of two ways. In some nations (often referred to as "automatic incorporation" states), treaties become binding as a art of domestic law as soon as formal ratification procedures have been adopted. In these nations, treaties are considered to be "self-executing" or capable of being given "direct effect" in domestic law such that courts and administrative agencies can construe the treaty directly and enforce the rights it grants to the owners of intellectual property products.

In other nations, however (often referred to as "legislative incorporation" states), treaties are considered to be "non-self-executing" and can only become binding in domestc law once the parliament or legislature has adopted legislation to implement the treaty. In these nations, owners of intellectual property products rely on this domestic legislation rather than on the treaties themselves when they seek to enforce rights granted to them under the treaties.

The territoriality of IPRs : Because of the limited scope of international IPR agreements, there are to this day (with the limited excption of the European Union) no *international* IPRs available to inventors and creators who seek to market their products across borders; rather, IPRs are territorial in nature and are acquired and enforced on a country-by-country basis under territorially-circumscribed *national* IPR laws.

Thus, for example, the inventor of a genetically enhanced variety of corn who seeks patent protection for that variety must apply for protection in each country in which he or she hopes to sell the corn. The inventor must comply with all of the requirements that each country imposes for granting patent rights to the new variety. Similarly, once protection is granted, issues such as the scope of the exclusive rights the inventor enjoys in the patented plant variety, the term of patent protection and the limitations imposed on the inventor's rights are all determined by the different national laws.

Recent international agreements have achieved some modest forms of procedural harmonization, but they have not altered the fundamental premise that national laws rather than international treaties are the immediate source of nearly all private rights in intellectual property products.

Two basic principles flow from this territorial approach to protecting IPRs. First, where national laws differ as to the scope or content of the protection they provide to intellectual property products, the rights enjoyed by the owners of those products will vary in different national jurisdictions. Second, territoriality implies that each nation has the right to decide on the form of IPR protection to be granted within its own borders, provided that it complies with the obligations contained in international IPR agreements to which it is a party.

Although territoriality thus gives governments some autonomy to set national IPR policies within their own borders, states often view the policies other governments choose as a subject of concern. Indeed, the global reach of markets for intellectual property products makes this concern a necessity.

To take just one example, most patent laws grant inventors owning patents within a state the right to prevent the importation into that state of products created in other nations that contain the patented invention. Thus, where distribution markets transcend national borders, an industry may find itself precluded from distributing products in other jurisdictions as a result of patent rights, as occurred when Indian cotton producers were precluded from importing certain forms of transgenic cotton into the United States.

Core obligations imposed by international intellectual property agreements : The territorial approach to IPR protection appears at first to present myriad difficulties for creators and owners of intellectual property products. In fact, however, the content of each nation's IPR laws are often quite similar since they have been shaped by international IPR agreements ratified by many states. In addition, the obligations these agreements impose have expanded over time, thus narrowing (although by no means eliminating) the differences among national intellectual property systems. The following sections briefly explain the core obligations contained in most international IPR agreements. More detailed information concerning the rights and obligations contained in specific IPR agreements relating to plant varieties is provided in Part II below.

National treatment : One of the cornerstones of international IPR agreements is the national treatment principle. National treatment bars discrimination against foreign IPR owners by requiring that each state provide the same IPRs to private parties from other member states as are provided to the state's own nationals. National treatment levels the playing field among treaty parties and prevents a state from giving its own creators and inventors unfair advantages over foreign creators and inventors. In the absence of national treatment, for example, domestic firms could freely exploit intellectual property products created in other member states while simultaneously enjoying legal protection within their own domestic markets.

Reciprocity : The provisions of several intellectual property treaties contain a limited exception to national treatment known as reciprocity. Where a treaty permits reciprocity, State A may condition the grant of legal protection to intellectual property products from State B upon State B's granting of legal protection to intellectual property products from State A. Reciprocity is often applied to new IPRs as means of encouraging other nations to recognize the new rights and extend their protection to foreign nationals. Once a large number of states have recognized the new IPR, they may revise the treaty to eliminate the reciprocity option and impose a national treatment obligation.

Most favoured nation treatment : The most favoured nation ("MFN") principle is a common feature of international trade agreements but has only recently been applied to IPRs. The principle extends the national treatment rule by compelling a government that provides a privilege or benefit to *one* state within a treaty system automatically to grant that same privilege or benefit to *all* states within the same system. The MFN principle thus prevents a subset of states within a larger treaty system from entering into bilateral or other special agreements among themselves, unless they grant the rights contained in those agreements to all other parties within the larger treaty system.

Subject matter and eligibility requirements : Intellectual property agreements specify the subject matter characteristics of the products that are eligible for legal protection. In the context of patents, for example, a treaty may specify the types of inventions (such as products and processes) to which states parties must grant legal rights. These subject matter requirements are generally drafted using language that instructs member states concerning the basic characteristics that a product must possess for it to merit protection under domestic IPR laws, while preserving sufficient flexibility for states to tailor the details of protection to the particularities of their national legal systems.

Exclusive rights : Where an intellectual property product satisfies a treaty's subject matter eligibility requirements, states parties are required to grant an enumerated set of exclusive rights with respect to that product. These exclusive rights grant to the product's owner the power to exclude all third parties from engaging in the activity that the right covers (such as reproducing or modifying the product or distributing it to others). It is the exclusivity of the rights granted that allows IPR owners to recoup the investment of time, money and resources required to create intellectual property products. The particular exclusive rights mandated by IPR agreements differ depending on the specific type of IPR product at issue. They are drafted with greater or lesser degrees of specificity depending on the degree of consensus among member states' national IPR laws, thus affording varying levels of discretion to governments to implement the rights in their domestic legal systems.

Terms of protection and the public domain : Intellectual property agreements also specify for their states parties the minimum term of years during which intellectual property products must receive legal protection. Once that term has expired, the treaties do not require states to grant legal protection to the products. Thus, unless the state adopts a longer term of protection, after the expiration of the initial term of protection the product may be freely used by anyone for any purpose, including as a source for creating new products or simply for consumption. A corollary of this rule is that national IPR laws do not permit putative inventors and creators to claim IPRs in materials as they are found in nature or where they are already part of the public domain.

Exceptions and limitations to exclusive rights : International IPR agreements constrain the ability of national governments seeking to restrict the exercise of IPRs to achieve competing social or policy objectives, such as access to information, research, education and cultural development. Restrictions designed to achieve these objectives are generally known as "exceptions and limitations" to exclusive IPRs. These exceptions and limitations generally appear in two forms.

The first form permits third parties to engage in specified uses of intellectual property products without the permission of the rights holder and without the payment of remuneration. The second form is known as a "compulsory licence." Compulsory licences allow third parties to use intellectual property products without the owner's permission, but only upon the payment of adequate compensation. To prevent both forms of exceptions and limitations from eviscerating IPRs altogether, intellectual property agreements impose specific constraints the ability of member states to adopt them.

Enforcement provisions : The grant of IPRs in national laws would be meaningless without adequate and effective mechanisms to enforce those rights. For this reason, recent intellectual property agreements specify the types of enforcement provisions that member states must adopt in their national laws. These provisions include the imposition of civil and criminal penalties against any person who engages in acts of exploitation reserved to the owner of an intellectual property product without the owner's authorization. The penalties include civil judicial proceedings for monetary damages or an injunction to prevent the continued unauthorized use of the product and criminal proceedings commenced by the government itself.

The "Minimum Standards" Framework of Multilateral IPR Agreements

Taken together, these core provisions of international IPR agreements impose significant legal obligations on member states. The agreements do not, however, purport to definitively address all of the issues raised by the grant of legal protection to intellectual property products. For this reason, the treaties are often referred to as "minimum standards" agreements in that they create only a basic floor of legal protection to which all member states must adhere.

There are three important consequences of this minimum standards framework. First, it allows member states the discretion to interpret and apply those provisions of the treaties that are ambiguous or that reasonably permit more than one construction. Second, a minimum standards approach permits, but does not require, states to grant additional IPR protections within their national laws. And third and perhaps most importantly, the framework leaves member states free to enact laws that serve other political, economic or social objectives, even where those objectives are in tension with IPRs, provided that those laws are not inconsistent with the terms of IPR agreements.

In the last five years, however, an increasing number of developing countries have entered into bilateral or regional trade and investment treaties with the United States or with the European Community. These treaties often contain higher standards of intellectual property protection than those specified in multilateral intellectual property agreements. (Drahos, 2001; GRAIN, 2001) They thus restrict the minimum standards approach contained in the multilateral agreements, for example, by requiring states to enact specific intellectual property protection rules in their national laws or by requiring developing countries to follow the standards contained in multilateral agreements to which they are not parties.

The minimum standards framework (as limited by bilateral or regional treaties requiring higher standards of protection) provides a methodology for analysis but not a solution for every potential clash between IPRs and other governmental policies. This is particularly true for plant varieties and plant breeders' rights, an area of intellectual property protection regulated by several international IPR agreements and subject to diverse standards of legal protection under different domestic laws. To understand how national governments can reconcile competing and sometimes conflicting domestic laws and objectives consistently with the obligations imposed by different international IPR agreements, it is first necessary to examine other international agreements and international institutions relating to plant genetic resources that have promulgated policies in tension with IPRs.

Identifying the relevant international agreements and institutions relating to intellectual property rights in plant varieties.

Before turning to a more detailed discussion of policy objectives, it is important first to identify the principal international institutions and international agreements that generate legal rules and standards relating to IPRs in plant varieties and plant genetic resources, as well as the institutions and agreements that generate rules and standards in tension with IPRs. Although a comprehensive discussion of these institutions and agreements is beyond the scope of this study, a basic familiarity with their most important elements is necessary to understand the legal regime of intellectual property rights in plant genetic resources for food and agriculture.

World Intellectual Property Organization ("WIPO")

The World Intellectual Property Organization ("WIPO") is a specialized agency of the United Nations charged with "promot[ing] the protection of intellectual property throughout the world." (WIPO Convention, art. 3(i)). The WIPO Secretariat undertakes wide variety of activities relating to IPRs, including hosting diplomatic conferences of government representatives seeking to negotiate new international treaties. WIPO's staff also provide technical assistance and trainin to member states and their national intellectual property offices, epecially in developing countres. More recently, WIPO has created standing, expert and intergovernmental committees that conduct studies on particular intellectual propert topics and generate nonbinding guidelines and recommendations for consideration by WIPO members.

WIPO's recent activities in the area of plant genetic resources have been considerable. Issues relating to the intersecton of IPRs and PGRs have been rised during the negotiation of two multilateral patent agreements (the Patent Law Treaty and the Substantive Patent Law Treaty) and have been a principal subject of discussion and study in the recently created Intergovernmental Committee on Intellectual Property and Genetic Resources, Traditional Knowledge and Folklore ("IGC"). Developing states first sought to raise issues relating to the intersection of IPRs and PGRs during the WIPO-sponsored negotiation of the Patent Law Treaty in 1999.

They proposed the addition of an article in the treaty requiring applicants for inventions derived from genetic resources to demonstrate that they had received permission to access those resources from the country of origin. Industrialized countries opposed the proposal, arguing that it addressed substantive law issues that were inappropriate for inclusion in a treaty largely devoted to procedural issues. As a compromise, the WIPO Secretariat proposed the creationof a new intergovernmental committee (the IGC) to study the intellectual property aspects of genetic resources and traditional knowledge.

During the IGC's first five sessions between April 2001 and July 2003, WIPO members have supported a wide-ranging work programme for the

Committee, including: (1) creating a searchable database of clauses in contracts that regulate access to genetic resources or require benefit sharing; (2) studying technical issues raised by the disclosure of biodiversity-related information in patent applications; (3) creating databases of traditional knowledge; (4) identifying ways to document genetic resources and traditional knowledge in the public domain; and (5) studying the appropriate legal rules to protect traditional knowledge. Most recently, the WIPO General Assembly extended the IGC's mandate and authorized it to accelerate its work, which may include the development of new international agreements.

World Trade Organization ("WTO") : The WTO is a global intergovernmental organization dealing with the rules of trade between nations. It was established in 1994 at the conclusion of the Uruguay Round of trade negotiations held under the auspices of the General Agreement on Tariffs and Trade ("GATT"). As of July 2004, 147 states and customs territories ("WTO Members" or "Members") had joined the organization.

The WTO's substantive obligations are contained in a series of treaties relating to international trade, including agreements on trade in goods, services, agriculture, sanitary and phytosanitary measures, trade-related investment measures, technical barriers to trade, textiles and clothing and intellectual property rights. These treaties are linked together as annexes to the Agreement Establishing the WTO. Disputes between WTO Members relating to any of these treaties are to be resolved by the WTO Dispute Settlement Body, which is composed of *ad hoc* dispute settlement panels and a standing Appellate Body of trade experts. A WTO Member whose national laws or practices are challenged by another WTO Member and found to be incompatible with its WTO treaty obligations by the Dispute Settlement Body must modify those laws or practices or face the prospect of trade sanctions.

A more detailed discussion of the dispute settlement process in the context of the Agreement on Trade Related Aspects of Intellectual Property Rights ("TRIPs" or "TRIPs Agreement"), is provided in paragraph 2.3.1.3.3below. TRIPs is the WTO agreement that addresses IPRs. It is discussed in detail in section 2.3 below.

Convention on Biological Diversity ("CBD") : The CBD was opened for signature in 1992 and entered into force in 1993. As of July 2004, 188 states had ratified this agreement. The CBD's main objectives are the conservation of biological diversity, the sustainable use of its components, the fair and equitable sharing of benefits arising out of the utilization of genetic resources and the preservation of indigenous knowlege. (art. 1) The CBD also recognizes that nation states have the sovereign right to exploit their own resources and the authority to determine the conditions of access to them. (arts. 3 and 15) One of the mechanisms by which the CBD achieves its objectives is *in situ* conversation of plant genetic resources. Conservation *in situ* involves the preservation of ecosystems and natural habitats and the maintenance of viable populations of species in those settings. Such conservation occurs, for example, where farmers and indigenous communities safeguard traditional plant varieties in the locations where they grow naturally or are cultivated.

Although the CBD does not expressly refer to any international IPR agreements, it contains numerous provisions relating to IPRs, principally in article 16. In particular, article 16(5) recognizes that IPRs "may have an influence on the implementation" of the CBD. The article obliges member states to cooperate in order to ensure that IPRs are "supportive of and do not run counter to" the treaty's objectives. Other provisions make clear that the CBD is to be interpreted so as to preserve the rights of IPR owners recognized in international law. For example, articles 16(2)-(4) state that the transfer of technology and measures taken to gain access to such technology shall be consistent with the adequate and effective protection of IPRs recognized in international law. Thus, for example, where a government encourages foreign direct investment in industrial technologies (such as a biotechnological process used to insert new genetic sequences into existing plant varieties), it must respect any patent rights that the owner of that technology has acquired to protect it.

Over time, the biodiversity regime's approach to intellectual property protection has evolved beyond the text of the CBD. The Conference of the Parties (COP) - the convocation of CBD members that determines how the Convention should be applied and implemented - has given detailed attention to harmonizing IPRs with the CBD's objectives. In particular, developing countries active in the COP, together with the support of nongovernmental organizations, have expressed concern about the adverse effects of IPR and have sought to harness intellectual property rules to promote compliance with the Convention.

In response to this concern, official COP statements have stressed the need to "promot[e] increased mutual supportiveness and integration of biological diversity concerns and the protection of intellectual property rights." (CBD Decision IV/15, para. 9) Members of the COP have worked toward this objective by gathering information, commissioning case studies, holding workshops anddrafting guidelines and recommendations. For example, in April 2002, the COP adopted the "Bonn Guidelines on Access to Genetic Resources and Fair and Equitable Sharing of the Benefits Arising Out of their Utilization." (CBD Decsion VI/24, App. II, Annex, Part C) The Guidelines' most important recommendation encourages applicants for IPRs such as patents or plant breeders rights to disclose the country of origin of the genetic resources or the traditional knowledge upon which those IPRs are based. The Guidelines advocate these disclosures to monitor whether applicants for IPRs have obtained the prior informed consent of the country of origin and complied with the conditions of access (if any) that that country has adopted. (Id.)

Although the COP members adopted the Bonn Guidelines by consensus, the Guidelines' reliance upon intellectual property laws to promote compliance with the obligations of the CBD remains controversial. For example, a 2003 Communication from the European Commission to the European Parliament on the EC's implementation of the Bonn Guidelines considers whether to adopt "a self-standing disclosure requirement" for patent and plant variety protection applicants. (EC Implementation of Bonn Guidelines, COM (2003) 821) If adopted, such a disclosure requirement would not, however, be incorporated into national or European intellectual property laws. To the contrary, the failure to comply with the new disclosure requirement would, according to the EC Communication, "only have consequences outside the field of patent law." However, the EC Communication also states that the EC and its member states "should be ready to discuss, in the relevant international fora, the possibility of introducing *under intellectual property law* the same disclosure requirement... as a formal condition for patentability and not only as a self-standing obligation."

The Consultative Group on International Agricultural Research ("CGIAR"): CGIAR is an informal association of public and private donors founded in 1971 that supports an international network of agricultural research centres, each with its own governing body. CGIAR's mission is to engage in research, in partnership with other public and private entities, to promote sustainable agriculture in developing nations. With respect to plant genetic resources, CGIAR's principal method for achieving this objective is *ex situ* conversation. Networks of gene banks within CGIAR store and conserve seeds and propagating materials outside of their natural habitat for future use by farmers, researchers and breeders. The CGIAR network holds the world's largest *ex situ* collection of plant genetic resources for food and agriculture.

International Plant Genetic Resources Institute ("IPGRI") : IPGRI is the world's largest international institute dedicated to the conservation and use of plant genetic resources. Founded in 1974, IPGRI focuses on conservation, management and preservation of the diversity of plant genetic resources, through domestic, regional and international programmes and research initiatives.

The International Undertaking on Plant Genetic Resources (the "Undertaking") : FAO has helped to generate several nonbinding inernational instruments relating to plant genetic resources. The Undertaking, the first of these instruments, was adopted in 1983. As of 2000, 113 states were signatories to the Undertaking, thus pledging themselves to implement he recommendations it contains. For many years, the Undertaking served as the central legal instrument in FAO's global system for plant geneti resources, a system that includes a fund for the equitable sharing of benefits and a mechanism to give early warning about genetic resources under threat.

The Undertaking's principal objectives are to ensure that the need for conservation is globally recognized and that sufficient funds for this purpose are made available; to assist farmers and farming communities in the protection and conservation of PGRs and of the natural biosphere; and to allow farmers, their communities and countries to participate fully in the benefits derived from improved uses of PGRs, including through plant breeding. (WT/CTE/125, para. 11)

In its initial formulation, te Undertaking challenged a private property rights approach to plant genetic resources bydeclaring that all such resources, whether as cultivated by farmers in the field or modified through breeder innovations, were part of the "common heritage of mankind and consequently should be available without restriction." (art. 1). An intrpretation issued adopted by FAO in 1989, however, clarfied that plant breeders' rights were not incompatible with the Undertaking. It also recognized the interrelationship beteen the rights of traditional farmers (whose practice of saving seeds provided the raw genetic materials for innovation) and the rights of plant breeders (who use technology to achieve that innovation).

The International Treaty on Plant Genetic Resources for Food and Agriculture ("ITPGR") : On 3 November 2001, an intergovernmental conference sponsored by FAO adopted the text of a legally binding international agreement on plant genetic resources. The ITPGR entered into force on 29 June 2004. As of July of that year, the treaty had been ratified by 55 states and signed by an additional 50 countries. The ITPGR not only codifies and updates the nonbinding principles set out in the Undertaking and its subsequent revisions, but also contains provisions relevant to IPRs in plant genetic resources and plant varieties. Because of its importance to the subject of this study, an extended discussion of the ITPGR appears in section 4.3 below.

Specific Policy Objectives in Tension with IPRs

Having identified the principal institutions and agreements relating to plant genetic resources, the next sections address specific critiques of IPRs as applied to plants and plant varieties and the policy arguments that inform those critiques.

Preserving genetic diversity : Granting IPRs to plant breeders has uncertain consequences for preserving plant genetic diversity. Given the incentives that IPRs create for private parties to invest in research and breeding techniques relating to new plant varieties, it might be thought that

IPRs would lead to an increase in plant genetic diversity over time. A number of commentators have argued, however, that diversity is eroded rather than enhanced by granting IPRs to plant breeders. According to this iew, *in situ* conservation by indigenous farmers diminished as they began to rely on commercial plant breeders for seeds and other propagating material. Rather than using informal breeding techiques to experiment with the creation of new varieties suitable for local growing conditions, indigenous farmers came to depend upon third party plant breeders to provide them with seeds possessing uniform genetic characteristics. The plant varieties that have come to dominate agriculture as a result of this dependence may possess many beneficial characteristics, but they do not enjoy the adaptive abilities of less well known and informally bred varieties.

To evaluate these competing assertions about the effects of IPRs on plant genetic diversity, empirical testing would be useful. Unfortunately, empirical studies of the causal impact of IPRs on plant genetic diversity are difficult to devise because of the uncertainty of screening out other relevant causal variables. Empirical studies that do exist often focus on industrialized countries rather than developing states and tend to rely on anecdotal evidence. As a result, the broader applicability of their conclusions is uncertain and often controversial.

Farmers' rights : A second challenge to IPRs concerns the relationship between farmers' rights and IPRs in plant varieties. The concept of farmers' rights was developed to reflect the contributions that traditional farmers, particularly in the developing world, have made to the preservation and improvement of plant genetic resources. FAO Resolution 5/89 defines farmers' rights as "rights arising from the past, present and future contributions of farmers in conserving, improving and making available plant genetic resources, particularly those in centres of origin/diversity." Such rights are also recognized in article 9 of the ITPGR.

Farmers' rights are in tension with IPRs for plant breeders because many farmers and farming communities do not claim exclusive rights in the cultivated landraces and plant varieties they have cultivated over time. Moreover, the subject matter requirements of existing IPR laws, which as explained below are designed to protect innovations in new and clearly distinguishable plant *varieties*, often cannot accommodate contributions of individual farmers using more informal methods to select for better crops or sought-after plant *characteristics*.

Advocates of farmers' rights have developed different approaches to address this situation and to reward farmers for their contributions to plant genetic diversity. The first approach involves situating the traditional practices of farmers as exceptions to the exclusive rights of plant breeders under existing

IPR laws. In other words, breeders are precluded from demanding payment from farmers who engage in certain farming practices, such as saving seeds and planting seeds saved from prior purchases, or informally exchanging seeds. A second approach seeks to modify existing IPR laws so as to permit farmers themselves to claim exclusive rights in the plant varieties they cultivate informally. A third approach involves recognizing farmers' rights not through IPRs but through benefit sharing mechanisms, such as financial payments and technology transfers, that compensate farmers for their contributions to plant genetic diversity. This last approach - which is recognized in article 9 of the ITPGR- questions whether farmers in fact have "rights" as that term is understood within an intellectual property paradigm, while acknowledging the need to reward their contributions to plant genetic diversity.

Traditional knowledge: An issue closely related to farmers' rights is the recognition and protection of the plant-related knowledge, innovations and practices of indigenous and local communities. Advocates assert that those claiming IPRs in plant genetic resources and plant varieties often utilize such knowledge without adequately acknowledging the contributions of the communities who possess it. Mechanisms to redress this problem are similar to those discussed above relating to farmers' rights, a linkage reflected in article 9 of the ITPGR. They include modifying existing IPR laws to recognize traditional knowledge as a form of intellectual property and protecting traditional knowledge through *sui generis* national laws that recognize the communal ownership of much of traditional knowledge. In the fall of 2000, the World Intellectual Property Organization (WIPO) established a new Intergovernmental Committee on Intellectual Property and Genetic Resources, Traditional Knowledge and Folklore to study these issues.

Regulating access to plant genetic resources : Plant breeders and others seeking to develop plant-related innovations need access to existing stocks of plant germplasm for breeding, research and development. Access issues arise for both *in situ* and *ex situ* collections of seeds and plant propagating material.

With respect to *in situ* collections, a number of states have enacted national access laws and regulations to control third party access. These access controls are consistent with the CBD, which grants states the sovereign right to control their plant genetic resources. They also further the CBD's benefit sharing objectives by conditioning access upon the current or future payment of compensation or the transfer of technology to the state providing access. The specific conditions of access are often specified in Material Transfer Agreements ("MTAs") entered into between the entity (whether governmental or private) that owns or controls the germplasm and the entity (usually private) that seeks access to it.

Similar access issues arise with regard to *ex situ* collections of plant germplasm, including the seed banks maintained under the auspices of CGIAR. Although pursuant to a 1994 agreement between CGIAR and FAO the autonomous agricultural research centres under CGIAR's auspices each create their own access regulations, certain designated germplasm is to be held by the centres "in trust for the benefit of the international community." (art. 3(a)) With respect to IPRs, the 1994 agreement provides that a centre "shall not claim legal ownership over the designed germplasm, nor shall it seek any [IPRs] over that germplasm or related information." (art. 3(b)). The centres also commit to apply the same restrictions on IPRs in MTAs with third parties to whom the centre transfers designated germplasm. (art. 10)

"Biopiracy" and property rights in unimproved plant materials : Issues of access to plant genetic resources are linked to larger controversies concerning the propriety of granting IPRs in the raw materials accessed.

Under settled principles of intellectual property law, unimproved plant germplasm already in the public domain cannot be removed and privatized. Nevertheless, there are several reported cases and one empirical study which indicate that plant breeders and commercial entities have on occasion been granted IPRs in wild plant varieties or in germplasm found in CGIAR seed collections.

It is unclear whether such rights were granted because of inadvertence, insufficient or inaccurate information provided by the IPR applicant or particularities in national laws that limit the sources to which granting authorities may refer in determining what materials are in the public domain. Where such rights have been granted, however, states and interested NGOs have often succeeded in petitioning the authorities in other nations to remove legal protections they had previously granted.

Attempts to claim IPRs in unimproved plant gnetic resources have often been labelled as a form of "biopiracy." Biopiracy is not a legal term of art, however and has been loosely used to refer to any act by which a commercial entity seeks to obtain IPRs over biological resources, including plant varieties, that are seen as "belonging" to developing states or indigenous communities. Thus, even where an IPR claim relates to improvements to raw plant materials, certain governments and NGOs have labelled the entity seeking legal protection as a biopirate if it has not provided a fair return to those who granted access to the raw materials.

Plant breeders' research interests : Even as between groups of plant breeders, the scope of IPRs in plant varieties can be controversial. Tensions arise between first generation breeders who have secured legal protection for new varieties and second generation breeders who seek to utilize those new varieties to develop still more varieties. As with farmers' rights, it is possible to use the exceptions and limitations provisions of national IPR laws to permit second generation innovators to engage in such activities without the authorization of first generation breeders.

International IPR Agreements Regulating Plant Varieties and Plant Breeders' Rights

This section identifies and discusses the provisions of three different international IPR agreements that protect plant varieties and plant breeders' rights. The two major treaty systems that regulate these issues are the agreements established under the auspices of the *Union internationale pour la protection des obtentions vegetales* ("UPOV"), and the TRIPs Agreement included within the family of treaties administered by the World Trade Organization ("WTO"). These two treaty systems each contain a comprehensive set of rules for their members regarding IPRs over plant varieties.

In short, the UPOV treaties adopt a *sui generis* system of protection (that is, a system that is unique, or of its own kind) especially tailored to the needs of plant breeders. The TRIPs Agreement requires WTO Members to protect new plant varieties using patent rights, a *sui generis* system or some combination thereof. Because TRIPs provides states with this flexibility and because the treaty has an uncertain relationship to the previously adopted UPOV conventions, national governments face a wide array of options in choosing the intellectual property regime applicable to plant varieties. This section of the study outlines the requirements imposed by these two treaty systems, and Part III of the study then identifies and analyses these options.

The UPOV Acts

The first UPOV Act was drafted in 1961, principally by industrialized governments seeking to provide potections for plant breeders in their own and overseas markets. The UPOV was later revised in Acts adopted in 1972, 1978 and 1991.

As of July 2004, 55 stats were parties to UPOV - 29 to the 1978 Act, 24 to the 1991 Act and 2 to the 1972 Act. (States Parties to the UPOV, Status on 30 June 2004). As a result, this study will focus on the two most recent UPOV Ats. Many of the accessions to these Acts are quite recent, with a number of developing states and countries in transition to a market economy acceding in the 1990s. The 1991 Act entered into force on 24 April 1998 and on that same date the 1978 Act was closed to future accessions except by a few states already in the process of adhering to it.

As explained in section 1.3.1 above, countries generally give domestic effect to the UPOV Act to which they are a party in one of two ways. In "automatic incorporation" states, courts and administrative agencies directly apply and enforce the Act, although implementing legislation is often needed to authorize administrative agencies to process applications to protect new plant varieties. In "legislative incorporation" states, by contrast, the UPOV Act does not become enforceable in domestic law until the state enacts a national plant variety protection law that conforms to the Act's requirements.

The 1978 UPOV Act

The 1978 UPOV Act adopts most of the international IPR obligations set out in Part I above, including a definition of the applicable subject matter and protected material, eligibility requirements, exclusive rights, national treatment, reciprocity, terms of protection and exceptions and limitations to exclusive rights. It does not, however, contain any provisions on MFN treatment or enforcement.

Subject Matter Requirements

Limited number of protected genera or species : Not all plant varieties must be protected under the 1978 Act. Rather, article 4 provides that member states are to progressively extend protection to an increasing number of genera or species, beginning with five on the date the treaty enters into force for that state and ending with twenty-four within eight years.

In addition, member states are free to limit the Act's application within a particular genus or species to varieties with a particular manner of reproduction or multiplication, or a certain end-use.

Preclusion of dual protection with breeder's right and patent : The 1978 Act permits its signatories to protect plant varieties either with a distinct breeder's right or with a patent. However, article 2(1) precludes member states from granting both forms of protection "for one and the same botanical genus or species."

Protection of discovered varieties required : Although the 1978 Act focuses on plant varieties created through classical breeding methods, it is generally accepted that the treaty requires member states to protect varieties which have been discovered. This has beeninferred from article 6.1(a) which indicates that a protected variety may result from a natural source of initial variation.

Eligibility Requirements

Assuming that a plant variety falls within a protected genera or species, it is eligible for protection under the 1978 Act only if it is: (1) new, (2) distinct from existing or commonly known varieties, (3) homogenous or uniform and (4) stable. (art. 6) When a variety fulfils these four criteria, it is listed in a national register or catalogue which publicly discloses that the variety is protected. (FIS/ASSINSEL, 2001a).

Novelty : To avoid protection for plant varieties that have already been exploited or are a matter of common knowledge, a new variety in which a breeder seeks protection must not have been sold on the market for more than a specified period of years prior to the date of application for protection. The 1978 Act specifies the maximum number of years during which such pre-application sales have occurred, with different periods of time set for different types of plants as well as for sales within the territory of the applicant state versus the territory of other states. (art. 6(1)(b)).

Distinctness : The 1978 Act states that a protectable plant variety must be "clearly distinguishable in one or more important characteristics from any other variety whose existence is a matter of common knowledge at the time when protection is applied for." (art. 6(1)(a)).

Although the treaty itself does not further define distinctness, the Guidelines for the Conduct of Tests for Distinctness, Homogeneity and Stability (UPOV Guidelines) use both qualitative and quantitative plant characteristics, including such visible attributes as leaf shape, stem length and colour, to determine if the difference between varieties is "clear and consistent." As explained below in the analysis of the UPOV 1991 Act, the concept of distinctness is critical to determining the scope of a breeder's right in plants that are closely related but not identical to a protected variety.

Homogeneity : Under the 1978 UPOV Act, a variety has to be "sufficiently homogeneous, having regard to the particular features of its sexual reproduction or vegetative propagation." (art. 6(1)(c)). The UPOV Guidelines further clarify that to be considered homogeneous, the variation shown by a variety must be "as limited as necessary to permit accurate description and assessment of distinctness and to ensure stability." The homogeneity requirement has been criticized by commentators as discouraging variability in plant varieties that are often useful for sound agricultural practices and as denying protection to breeders of cultivated landraces that exhibit diversity traits.

It is thus one of the sources for the criticisms identified in paragraph 1.3.6.1 above that plant breeders' rights are reducing plant genetic diversity by rewarding breeders of uniform plant varieties.

Stability : The stability requirement is a temporal one, requiring the breeder to show that the essential characteristics of its variety are homogeneous or uniform over time, even after repeated reproduction or propagation. (art. 6(1)(d)) In practice, what has been shown to be homogeneous is usually considered to be stable as well. For this reason, the stability requirement has engendered the same sort of critiques as the uniformity requirement in its preclusion of protection for cultivated landraces and other traditional plant varieties.

Protected material : The 1978 Act requires its signatories to protect a variety's reproductive or vegetative propagating material. The Act does not require protection of harvested material, with the exception of ornamental plants that are used for commercial propagating purposes. (art. 5(1))

Breeders' exclusive rights : Under article 5 of the 1978 Act, any person seeking to engage in the following three acts, with respect to a protected variety's reproductive or vegetative propagating material, must obtain the prior authorization of the breeder: (1) production for purposes of commercial marketing, (2) the offering for sale and (3) marketing. The 1978 Act does not, however, require member states to extend these exclusive rights to harvested material or other marketed products.

National treatment and reciprocity : Member states must grant the three exclusive rights referred to in the preceding paragraph in the same manner to both national breeders and to breeders who reside in or are nationals of other 1978 Act member states. However, where a state extends legal protection to a specific genus or species, or where it provides more extensive exclusive rights to breeders than the rights required under the treaty, reciprocity is permitted. Thus, a state providing these additional rights may restrict protection to breeders from those member states that apply the Act to the same genus or species, or that provide such additional exclusive rights to their own nationals. (arts. 3 and 5(4))

Term of protection : The 1978 Act requires a minimum term of protection of fifteen years, with the exception of vines, forest trees, fruit trees and ornamental trees, which are protected for no less than eighteen years.

Exceptions and limitatios : Two major exceptions and limitations to exclusive rights exist under the 1978 Act: (1) a breeders' exemption and (2) a farmers' privilege. The Act also permits members to impose compulsory licences.

Breeders' exemption: This exemption in article 5(3) precludes member states from granting to breeders of protected varieties the right to authorize or refrain from authorizing other breeders seeking to use the protected variety to create new varieties or to market those new varieties. States are permitted to grant breeders such an authorization right only if the repeated use of the protected variety is necessary for the commercial production of the new variety. According to the International Association of Plant Breeders and the International Seed Federation, this breeders' exemption "is essential for continued progress from plant breeding." (FIS/ASSINSEL, 1999)

Farmers' privilege : The focus of the 1978 Act on commercial exploitation of protected plant varieties has been interpreted to allow the use of seeds and propagating material for noncommercial purposes without the breeder's prior authorization. In national plant variety protection laws, this implicit noncommercial exception most frequently benefits farmers who purchase the seeds of protected varieties. The scope of this so-called farmers' privilege varies widely, however. Some nations only permit farmers to plant seeds saved from prior purchases to be used ontheir own land holdings, while others allow them not only to replant but also to sell limited quantities of seeds for reproductive purposes, a practice often referred to as "brown bagging."

Compulsory licences in the public interest : Article 9 of the 1978 Act permits members to restrict breeders' exclusive rights for "reasons of public interest." Where such restrictions are enacted to ensure the widespread distribution of the variety (such as where the breeder fails to supply the demand for the variety in a reasonable quantity and price or unreasonably refuses to license the variety to third parties), the breeder must receive equitable remuneration.

The 1991 UPOV Act

The limited scope of the 1978 Act led a number of member states of the UPOV to adopt a revised Act with enhaced rights for plant breeders. The major revisions o the 1991 Act are discussed below:

Subject Matter Requirements

Phased-in protection of all genera or species : Unlike its predecessor, the 1991 Act requires states to protect at least fifteen plant genera or species upon ratifying or acceding to the Act, and to extend protection to all plant varieties within ten years. (art. 3(2)) It also contains a definition of a plant "variety" as a "plant grouping within a single botanical taxon of the lowest known rank" which can be "defined by the expression of the characteristics resulting from a given genotype or combination of genotypes; distinguished from any other plant grouping by the expression of at least one of the said characteristics; and considered as a unit with regard to its suitability for being propagated unchanged." (art. 1(vi)) No definition of "variety" appears in the 1978 Act, which indicates that member states to that earlier treaty have greater discretion in defining the characteristics of plant groupings that qualify for protection.

Dual protection with breeders' right and patent permitted : In response to demands from breeders in industrialized counties, the 1991 Act removed the 1978 Act's ban on dual protection and now permits member states to protect the same plant variety with both a breeders' right and a patent.

Protection of discovered varieties : The 1991 Act makes explicit the 1978 Act's implicit requirement that discovered varieties be protected. It does so through article 1(iv)'s definition of a "breeder" as including a "person who bred, or discovered and developed, a variety."

Eligibility Requirements

The four eligibility requirements that must be demonstrated to merit protection for a specific variety - novelty, distinctness, uniformity and stability - are preserved in the 1991 Act, subject to only minor changes in scope and wording. (arts. 7-9) Thus, the 1991 Act has received the same criticism as the 1978 Act for its encouraging of genetic standardization and its inability to protect more diverse plant varieties, traditional varieties or cultivated landraces.

Breeders' Exclusive Rights in Protected Material

Extensive additions to the 1978 Act were made with respect to the exclusive rights enjoyed by breeders in protected material of plant varieties.

Enumerated exclusive rights in propagating material : The breeder's prior authorization must be obtained for the use of reproductive or vegetative propagating material of the variety for (1) production or reproduction, (2) conditioning for the purpose of propagation, (3) offering for sale, (4) selling or marketing, (5) exporting, (6) importing and (7) stocking for any of these purposes.

Extension of rights to harvested material and products : These exclusive rights apply not only to propagating material but also to harvested material, where the harvest has been obtained through an unauthorized use of the propagating material and the breeder has not had a reasonable opportunity to exercise his or her right in relation to that material. In addition, member states are permitted but not required to provide additional rights to breeders, including rights in products made directly from harvested material.

National Treatment Required

Whatever the particular exclusive rights member states adopt in their plant variety protection laws, those rights must also be provided to the nationals of other 1991 Act member states. Unlike the 1978 Act, granting rights only on condition of reciprocity is not permitted. (art. 4) For example, if a 1991 Act member chooses to grant more expansive exclusive rights to breeders than the exclusive rights required by the 1991 Act, it must grant those rights to breeders from all other 1991 Act member states. The state cannot choose to grant such expansive rights to only those 1991 Act states that also provide the same level of rights protection to the first state's nationals.

Terms of Protection

The 1991 Act extends the term of protection to 20 years, and requires a 25-year term for tree and vine varieties.

Exceptions and Limitations

A second area of major revision concerns the reduced scope of exceptions and limitations to breeders' exclusive rights, which is found in article 15 of the 1991 Act.

Private, noncommercial exception : The 1991 Act makes explicit what was only implicit in the 1978 Act, namely that private, noncommercial activities with respect to new varieties are outside of the breeder's control. This exception would presumably permit subsistence farmers to use protected seeds and other propagating material for their own consumption.

Research exception : The 1991 Act also recognizes that breeders cannot restrict "acts done for experimental purposes." This exception would permit research and testing of protected varieties for scientific purposes that does not lead to commercial exploitation.

Limited breeders' exemption : Like its predecessor, the 1991 Act recognizes the right of breeders to use protected varieties to create new varieties. However, this exception is itself restricted in its application to such new varieties as are not "essentially derived" from protected varieties. (arts. 14(5) and 15) The drafters added this restriction to prevent second generation breeders from making merely cosmetic changes to existing varieties in order to claim protection for a new variety.

The concept of essential derivation has proved highly controversial in practice, however. Breeders have been unable to agree on a definition of the minimum genetic distance required for second generation varieties to be treated as not essentially derived from an earlier variety and thus outside of the first breeder's control. (FIS/ASSINSEL, 1999). Nevertheless, the overall effect of this provision of the 1991 Act has been to narrow the breeders' exemption and expand the IPRs of first generation breeders.

Limited farmers' privilege : The 1991 Act recognizes an explicit farmers' privilege, but one that is limited in scope. In particular, each member state may enact a provision in its national plant variety protection laws permitting farmers to use for propagating purposes "on their own holdings" the product of the harvest which they obtained by planting a protected variety "on their own holdings." In addition, this privilege must be exercised "within reasonable limits and subject to the safeguarding of the legitimate interests of the breeder." (art. 15(2)) Several points can be inferred from this language.

First, unlike the 1978 Act, the 1991 version of the farmers' privilege does not authorize farmers to sell or exchange seeds with other farmers for propagating purposes, a limitation that commentators have criticized as inconsistent with the practices of farmers in many developing nations, where seeds are exchanged for purposes of crop and variety rotation.

Second, according to the International Association of Plant Breeders, the "reasonable limits" language in article 15(2) requires states to restrict the acreage, quantity of seed and species subject to the farmers' privilege, while the requirement to safeguard breeders' "legitimate interests" requires farmers to pay some form of remuneration to the breeder for their privileged acts. (FIS/ ASSINSEL, 2001a) The latter assertion is controversial, however, and has not been enacted in the national laws of all the 1991 Act member states. For example, the EC's Council Regulation on Community Plant Variety Rights, No. 2100/94 of 27 July 1994, requires such payments except in the case of small farmers, whereas the United States' Plant Varieties Protection Act of 1994, 7 U.S.C. Sec. 2543, does not.

Compulsory licences in the public interest : Article 17 of the 1991 Act contains a compulsory licence provision similar to that found in the 1978 Act. It permits members to restrict breeders' exclusive rights only for reasons of public interest and requires payment of equitable remuneration to the breeder whose rights are limited.

Resistance to the 1991 Act

Many developing nations, particularly those in Africa, have resisted ratifying the 1991 Act or adopting it as the standard for their plant variety protection laws. The foreign ministers of the more than 50-member Organization for African Unity (now the African Union) issued a statement at a January 1999 meeting calling for a hold on IPR protection for plant varieties until an Africa-wide system has been eveloped that grants greater recognition to the cultivation practices of indigenous communities.

However, at a subsequent meeting of the Organisation Africaine de la propriete Intellectuelle (OAPIO), patent officials from sixteen francophone African nations recommended that their countries adopt the 1991 Act. (Machipisa, 1999). As of the date of this study, Tunisia, Kenya and South Africa are the only African UPOV member states. Tunisia ratified the 1991 Act in August 2003, whereas the other two countries are parties to the 1978 Act. Other developing states parties to the 1978 Act are Bolivia, China, Ecuador, Kyrgyzstan, Mexico, Nicaragua, Panama, Paraguay, Trinidad and Tobago and Uruguay. These and other developing countries are, however, under pressure to ratify the 1991 Act as part of bilateral or regional trade and investment agreements.

The TRIPs Agreement

Although the UPOV Acts have provided IPR protection for plant varieties for more than forty years, their significance has recently been overshadowed by a different intellectual property treaty, the Agreement on Trade-Related Aspects of Intellectual Property Rights ("TRIPs" or the "TRIPs Agreement"). Adopted in 1994 as a treaty administered by the WTO, TRIPs is the first and only IPR treaty that seeks to establish universal, minimum standards of protection across the major fields of intellectual property, including patents, copyrights, trademarks, industrial designs, integrated circuits and trade secrets.

Although the TRIPs Agreement devotes only minimal attention to plant breeders' rights or plant variety protection and does not even mention the UPOV Acts, its adoption has done more to encourage the legal protection of plant varieties than any other international agreement.

TRIPs as a Spur to Plant Variety Protection

TRIPs' influence on plant variety protection stems from the following sources: (1) its link to other international trade agreements; (2) its widespread adherence by states in both the industrialized and developing world; (3) its novel enforcement, review and dispute settlement provisions; (4) the requirement in TRIPs article 27.3(b) that its signatories must provide protection for plant varieties "either by patents or by an effective *sui generis* system or by any combination thereof"; and (5) a formal review of article 27.3(b) which was scheduled to be held in 1999. The following paragraphs briefly address each of these five issues. Subsequent sections devote more detailed treatment to the protection of plant varieties with patents and to the elements necessary to create an "effective *sui generis* system."

The Relationship of TRIPs to the WTO and the International Trading System

Unlike all prior intellectual property treaties, TRIPs is not a free-standing agreement concerned solely with IPRs. Rather, TRIPs is linked to a larger family of trade-related agreements concerning subjects such as trade in goods and services, agriculture, textiles and health-related restrictions on imports. All of these agreements were adopted within the WTO during the Uruguay Round of trade negotiations held between 1988 and 1994. As such, TRIPs was part of a global "package deal."

Industrialized nations secured a commitment from developing nations to provide minimum stadards of effective legal protection to intellectual property products, and in exchange developing nations received a commitment from industrialized countries to open thei domestic markets to goods and other products manufactured in the developing world.

Widespread Adherence to TRIPs

The result of this global bargain was widespread adherence to all WTO Agreements, including the TRIPs Agreement. As of July 2004, 147 states or customs territories were obligated to comply with TRIPs by virtue of their membership in the WTO. In addition, many states (particularly those in the developing world) who became bound by TRIPs had not ratified earlier intellectual property treaties and were thus required to make significant changes to their national laws to bring them into compliance with the Agreement.

In acknowledgement of this fact, TRIPs contains a phase-in provision pursuant to which most of the treaty's substantive obligatons became binding for developed nations in 1996, but did not become binding on developing nations and nations in transition to market economies until 2000. Least developed nations are given the most leeway and are not required to implement the treaty's substantive obligations until 2006.

The Enforcement, Review and Dispute Settlement Provisions of TRIPs

In addition to its widespread adherence, the influence of the TRIPs Agreement can be traced to its unique provisions relating to the enforcement of IPRs within national laws, the review of those national laws by the TRIPs Council and the mechanism for settlement of disputes between states leading to rulings backed up by the threat of trade sanctions.

Enforcement provisions : Unlike prior IPR agreements, TRIPs does not only specify the minimum substantive requirements for various forms of intellectual property. In addition, it requires WTO Members to adopt "effective" provisions within their national laws to permit the owners of intellectual property products to enforce their rights against those who infringe them. (art. 41.1) These enforcement provisions include detailed judicial and administrative remedies, border measures and criminal procedures. (articles 41 to 61) To take just one example applied to plant varieties, a breeder whose new variety is sold commercially without permission must be able to bring a civil judicial action seeking an injunction to stop the conduct of the unauthorized seller and to recover damages from him or her.

Review provisions : Article 68 establishes a Council for TRIPs to "monitor the operation of th[e] Agreement and... Members' compliance with their obligations" under the treaty. Since its founding, one of the Council's principal functions has been to formally review the national laws, regulations and judicial decisions of WTO Members in each area of intellectual property covered by TRIPs.

Several goals are served by the TRIPs Council's review functions. First, the reviews create an incentive for governments to bring their national laws and practices into compliance with the Agreement so that they may present positive information to the Council. Second, the reviews identify for both the reviewed state and for other WTO Members areas of the law which may not be in full compliance with the treaty. And finally, TRIPs Council reviews provide a critical opportunity to publicize national laws and practices relating to IPRs which might otherwise be difficult to obtain.

In the area of plant varieties, the TRIPs Council has gathered and organized a considerable amount of useful information concerning national government practices. In December 1998 the Council prepared a detailed list of questions concerning plant variety protection. It asked WTO Members who were already obligated to protect plant varieties to respond to these questions by indicating the manner in which their national laws provide such protection, with other WTO Members requested to use their best efforts to furnish such information. The Council later received and published responses from 17 states and from the European Communities and its member states. (WTO Job No. 2689 of 7 May 1999). In 2001 and 2002, the Council reissued these questions and invited WTO Members that had not already done so to provide information on their plant variety protection practices. Additional responses were received from six states. Information distilled from some of these responses is discussed in section 3.3 below.

Dispute settlement provisions : Another significant innovation of TRIPs is its dispute settlement system. Although prior intellectual property agreements, including the UPOV Acts, contained provisions for filing of complaints against non-complying treaty parties before the International Court of Justice, no state ever invoked this dispute settlement option. It was widely believed that pursuing such a complaint would be perceived as an unfriendly act, would be time-consuming and costly and that states were unlikely to implement decisions of the court. The TRIPs Agreement removes each of these concerns by linking to a WTO Agreement known as the Dispute Settlement Understanding which contains a streamlined dispute settlement mechanism that is one of the most widely used and effective interstate dispute resolution systems in international law. Indeed, the mere existence of such a system creates strong incentives for WTO Members to bring their national laws into compliance with the WTO Agreements, including TRIPs, to avoid the possibility of dispute settlement proceedings.

These strong incentives notwithstanding, some countries may not comply. In such a case, where one WTO Member believes that another Member has failed to fulfill its obligations under the Agreement, it may initiate consultations under the Dispute Settlement Understanding with a view to resolving the dispute. If the parties fail to reach agreement, the complaining state may then request the WTO to convene a three-member panel of experts to review its allegations. Such panels generally issue a decision within six months. The panel's decision is adopted by the WTO's Dispute Settlement Body unless the losing party or parties elect to appeal the decision to the seven-member Appellate Body, a standing tribunal of seven trade eperts who are authorized to review the panel's findings and issue a decision within three months. If either the Appellate Body or an unreviewed panel decision concludes that a WTO Member has violated the TRIPs Agreement, the Dispute Settlement Body will recommend that the defending state bring its national laws into compliance with the treaty. If the state fails to do so, the complaining state may then commence an arbitral proceeding to specify the amount of compensation that the defending state must pay to remedy the violation, or in lieu of such compesation the complaining state may seek authorization to impose trade sanctions on the non-complying Member. As of July 2004, WTO Members (mostly developed nations) had commenced 24 dispute settlement proceedings against both developed and developing nations concerning a wide variety of intellectual property issues. (WTO Dispute Settlement: Index of Disputes Issues: 15 July 2004) Approximately one third of these disputes were resolved prior to a decision by a panel, with the defending state agreeing to modify its laws to bring them into compliance with the Agreement. In other cases, however, WTO Members complied only after a panel or the Appellate Body had issued a decision against them. And in a few instances, Members that have not modified their laws instead have negotiated settlements involving the payment of compensation to the complaining Member.

As of the date of this study, there have been no TRIPs dispute settlement proceedings relating to intellectual property protection for plant varieties. Whether such proceedings will be brought in the future depends on a number of variables, many of which are still uncertain.

Three interrelated reasons suggest that a dispute settlement proceeding concerning plant variety protection is a likely possibility: first, the significant difference of views among WTO Members over the scope of legal protection to be provided to plant varieties; second, the fact that developing country WTO Members first became obligated to protect plant varieties only in 2000; and third, the absence of plant variety protection in many developing countries' national laws prior to the 2000 deadline. Taken together, these variables suggest that the plant breeders of a developed WTO Member who learn that their protected varieties are being exploited without authorization in developing WTO Members due to inadequate national laws will pressure their governments to file a complaint against such states to compel them to adhere to their treaty obligations.

Other factors, however, suggest that a complaint over plant varieties is unlikely to be filed, particularly within the next five to ten years. First, the initiation of the Doha Round of trade negotiations in November 2001 (discussed in Part IV below) is likely to cause WTO Members to show restraint in their dispute settlement strategies while the multi-year negotiations are proceeding. This is especially so given that the Doha Round will likely provide WTO Members with the first meaningful opportunity to harmonize other international obligations (such as those set forth in the CBD and the ITPGR) with the plant variety and other IPR protection requirements of the TRIPs Agreement. Second, past practice has shown that WTO Members are most likely to file dispute settlement complaints where they are pressured to do so by the owners of intellectual property products. It is unclear whether plant breeders can make a sufficient showing that their overseas markets are being harmed by other states' failure to provide adequate legal protection for plant varieties. Third, WTO Members are generally reticent about filing complaints where their probability of success is unclear. Therefore, given the uncertain scope of protection for plant varieties currently required by TRIPs, if Members enact *some form* of plant variety protection in their national laws, other Members are unlikely to challenge those laws unless they can identify a clear violation of the Agreement. Nevertheless, the potential for dispute settlement proceedings (and the trade sanctions they can engender) will continue to provide strong incentives for states to enact plant variety protection laws.

The Unique Scope of Plant Variety Protection Required by TRIPs

Article 27.3(b) contains the only textual provisions of the TRIPs Agreement relating to plant variety protection. The article states in relevant part:

Members may also exclude from patentability: (b) *plants* and animals other than microorganisms; and essentially biological processes for the production of plants or animals other han non-biological and microbiological processes. However, *Members shall provide for the protection for plant varieties either by patents or by an effective* sui generis *system or by any combination thereof.* (emphasis added) The meaning of this article has been the subject of significant debate among both WTO Members and nongovernmental organizations (NGOs) with differing views over the propriety of IPR protection for plant varieties. A detailed review of that debate and an analysis of article 27.3(b) is provided below. For present purposes, two overarching points are worth noting.

First, TRIPs' provisions on plant varieties do not refer to or incorporate any preexisting intellectual property agreements, including the 1978 and 1991 UPOV Acts. This omission contrasts sharply with other fields of intellectual property, such as patents, copyrights and trademarks, for which TRIPs expressly requires WTO Members to comply with the standards of protection contained in preexisting IPR agreements, such as the Berne Convention for the Protection of Literary and Artistic Works and the Paris Convention for the Protection of Industrial Property. As a result of this omission, WTO Members are neither required to become members of UPOV nor to enact national laws consistent with either UPOV Act in order to comply with their obligations under TRIPs. Although the drafting history of TRIPs does not explain this markedly different treatment of plant varieties, it seems likely that compliance with UPOV was not required because so few WTO Members were party to UPOV and those who were could not agree upon which of its two most recent Acts should serve as the standard for protection.

Second, article 27.3(b) permits WTO Members to protect plant varieties using one of three distinct approaches: (1) patent law, (2) an effective *sui generis* system or (3) a combination of elements from both systems. Thus, unlike most other areas of intellectual property protected by TRIPs, article 27.3(b) expressly grants Members significant discretion to choose the manner in which they will protect plant varieties and it contemplates that that discretion may be exercised differently by different states.

This discretion and the opportunity for divergent outcomes it engenders has significant consequences. On the one hand, TRIPs' failure to incorporate and build upon the preexisting UPOV Acts may have "a deharmonizing effect," (Correa 1994a) with states within the UPOV system enacting one type of plant variety protection law and states outside of that system enacting a different kind of law (which may or may not resemble each other). This could create significant complexities and uncertainties for plant breeders seeking to market protected varieties in different jurisdictions. On the other hand, this sanctioned diversity of legal approaches allows WTO Members to balance the protection of plant breeders' rights against the other important and competing societal goals identified in Part I, many of which are found in other international agreements. Seen from this perspective, article 27.3(b) provides a much needed "safe space" for governments to harmonize conflicting norms and policies - a space that is lacking in other areas of the TRIPs Agreement.

The Review of TRIPs Article 27.3(b)

The last sentence of article 27.3(b) states that "[t]he provisions of this subparagraph shall be reviewed four years after the date of entry into force of the WTO Agreement." Inasmuch as the WTO Agreement entered into force on 1 January 1995, the review contemplated by this article should have been conducted in 1999. The review was commenced but was not concluded, largely as a result of disputes between industrialized and developing nations over the scope of the review process. Governments submitted additional information and proposals to the TRIPs Council in 2000 and 2001, but no formal action was taken. With the launching of the Doha Round of trade talks in November 2001, however, the review of article 27.3(b) recommenced in earnest.

In June 2002, eleven developing states submitted a detailed proposal to amend TRIPs to "prevent systematic conflicts with the CBD arising from the implementation of TRIPS." (WTO Doc. IP/C/W/356, para. 11) The proposal seeks to compel all WTO Members to require applicants for patents relating to biological materials and traditional knowledge to disclose certain information as a condition of obtaining legal protection. (Id., para. 10)

The European Communities and Switzerland responded to the developing states' proposal witha compromise. The EU response calls for the negotiation of "a self-standing disclosure requirement," which, while not functioning as a new eligibility criterion for patent protection, "would allow Members to keep track, at [the] global level, of all patent applications with regard to genetic resources for which they have granted access." In June 2003, developing countries rejected this compromise and "reaffirm[ed] and strengthen[ed] their demand for a strong disclosure of origin mechanism" within TRIPs that would require "not only detailed information about who provided the [genetic] materials or the [traditional] knowledge used [in patent applications], but also positive proof of benefit sharing and of prior informed consent." African countries also proposed a new "Decision on Traditional Knowledge" to be incorporated into TRIPs.

"TRIPs Plus" Bilateral and Regional Agreements

In the decade since TRIPs was first adopted, the United States and the European Community have negotiated a growing number of bilateral and regional trade and investment treaties with developing countries. Many of these treaties contain provisions concerning IPRs. Commentators have referred to these treaties by the appellation "TRIPs plus" because they (1) contain intellectual property protection standards more stringent than those found in TRIPs, (2) obligate developing countries to implement TRIPs before the end of its specified transition periods or (3) require such countries to accede to or conform to the requirements of other multilateral intellectual property agreements. As one study recently noted, "[t]o the extent that ["TRIPs plus" treaties] include these additional aspects, they are pushing harmonization forward at a pace that is greater than is apparently possible within the framework of the WTO." (OECD, 2003) In addition, once a developing state enacts this higher level of IPR protection in its national laws, TRIPs' MFN clause obligates that state to extend the same protection to *all other WTO Members*. In this way, bilateral agreements can be used to ratchet up the level of protection for IPRs beyond what is required in the TRIPs Agreement.

Several "TRIPs plus" treaties contain provisions relating to plant variety protection that go beyond the minimum level of intellectual property protection specified in TRIPs. For example, the recently concluded U.S.-Central America Free Trade Agreement (applicable to Costa Rica, El Salvador, Guatemala, Honduras, Nicaragua and possibly to the Dominican Republic), the 2002 U.S.-Chile free trade agreement, the 2000 U.S.-Jordan free trade agreement and the 2000 EU-Mexico free trade agreement, as well as certain Euro-Mediterranean Association agreements and the draft Free Trade Agreement of the Americas, all mandate UPOV as the appropriate mechanism to protect plant breeders' rights. (GRAIN, 2004b) The treaties also require these countries to ratify the 1991 UPOV Act within specified time frames. Such obligations limit the discretion enjoyed by these countries to tailor their national laws to achieve policy objectives in tension with IPRs.

Patent Protection for Plant Varieties Under TRIPs

The TRIPs Agreement mandates its signatories to provide patent protection for any inventions in all fields of technology, provided that the inventions are "new, involve an inventive step and are capable of industrial application." (art. 27(1)) However, with regard to plant-related inventions, TRIPs permits Members to exclude from patentability altogether "plants," "essentially biological processes for the production of plants" and "plant varieties." (art. 27.3(b)) Thus, as presently written, the TRIPs Agreement would permit WTO Members to decline to protect plant varieties with a patent (provided, as discussed in section 2.3.3 below, that they protect such varieties with an effective *sui generis* right).

Applicability of Patent Protection for Plant Varieties

For this reason, it may initially appear that governments can ignore patents as they consider how to protect plant-related innovations in their national legal systems. Nevertheless, a basic understanding of the patent provisions of the TRIPs Agreement and a comparison of those provisions with the plant breeders' rights approach contained in the UPOV Acts is essential for a number of reasons.

First, extending patent protection to plant-related inventions and innovations remains an option for national governments. Recall that TRIPs' "minimum standards" framework expressly contemplates that WTO Members may provide greater protection for IPRs than are mandated by the Agreement. Article 27.3(b) in particular invites Members to protect plant varieties with patents or with a combination of patents and a *sui generis* system. A number of governments in the industrialized world, including the United States, Japan, Australia, New Zealand, Sweden and the United Kingdom, have capitalized on this opportunity by permitting plant breeders to obtain patent protection in new varieties provided that the eligibility requirements for a patent have been met.

Second, because TRIPs does not require any patent protection for plant-related innovations, it follows as a matter of course that the treaty does not compel WTO Members to adopt any particular form of patent protection. This allows governments the option of including plant varieties within their existing utility patent statutes and/or of enacting a separate statute applicable exclusively to plants. (Utility patents are generally granted to "any new and useful process, machine, manufacture or composition of matter, or any new and useful improvement thereof... " In making this decision, however, it will be useful for governments to understand the traditional elements of patent law and the ways in which they may be varied within the confines of the TRIPs Agreement. The United States, for example, has enacted multiple plant variety protection laws.

It was one of the first nations to adopt, in the Plant Patent Act of 1930, a unique form of protection applicable to asexually reproducing plants. In addition, since the mid 1980s, U.S. courts and the U.S. Patent and Trademark Office have concluded that plant breeders may also seek standard utility patents under the U.S. Patent Act. In December 2001, the U.S. Supreme Court confirmed that breeders could apply for both forms of protection with respect to the same variety. Such a result is prohibited by the 1978 UPOV Act but permitted under the 1991 Act to which the U.S. is a party.

Third, an understanding of patent principles is also necessary because of the difficulty in some nations of distinguishing between plant varieties (which are often excluded from national or regional patent laws) and other plant-related innovations (to which patent protection may be extended). In Europe, for example, article 53(b) of the European Patent Convention prohibits the patenting of "plant varieties" as such. Nevertheless, the European Patent Office recently confirmed that claims to patent protection that are broadly drawn to encompass "plants" or an invention broader than a single variety *may* be patented, even though such claims may encompass multiple varieties. As a result, it may be possible for plant breeders in Europe to fashion their claims to receive *de facto* patent protection for new plant varieties.

There are significant differences in approach between plant breeders' rights and patents. In the case of plant breeders' rights, the eligibility requirements for protection are not onerous, but the scope of protection granted is quite narrow, both in terms of exclusive rights and the various exceptions and limitations to those rights. Patent laws strike a very different balance. Eligibility requirements are high and difficult to meet, but once granted a patent conveys broad rights to exclude third parties from exploiting the patented invention. Depending on the needs and level of development of plant breeder industries within its territory, a government may decide that either or both forms of protection will provide the appropriate incentives to encourage plant-related research and innovation.

Fifth, and finally, a review of the TRIPs Agreement's plant-related patent rules will soon occur in the context of the Doha Round of trade negotiations. Such a review may produce an agreement to amend TRIPs to require WTO Members to extend patent protection to plant-related innovations or plant varieties. National governments which have thus far eschewed patent protection would then need to revise their national laws to make such protection available, in addition to or instead of the legal protection provided under national plant variety protection laws.

At a minimum, therefore, governments interested in promoting plant-related innovations should be apprised of the relevant legal issues so that they may contribute in an informed way to the upcoming discussions.

Subject Matter Eligibility Requirements

According to the text of the TRIPs Agreement and authoritative commentary interpreting it, plant-related innovations may be patented if they are "inventions" that are novel, non-obvious and useful. Although commentators frequently assert that these cumulative requirements impose significant barriers to the patenting of plant varieties, in practice breeders have succeeded in patenting not only new varieties, but also hybrid plants and inbred and hybrid plant lines. As two commentators recently noted, since 1985 the U.S. Patent and Trademark Office "has granted hundreds of utility patents on all aspects of innovation relating to plant science: plants themselves, seeds, breeding methods and plant biotechnology."

Inventions : The TRIPs Agreement does not define the term "invention," thus leaving its precise meaning to national patent laws or regional patent rules such as those operating within the European Union.

All such laws recognize that an invention must be more than a mere discovery of a natural phenomenon or naturally occurring substance. Thus, for example, a breeder who merely identified the existence of a previously unknown wild variety or a cultivated landrace that was known only to an isolated indigenous community should not be permitted to patent that variety or landrace.

The trend among the industrialized countries, however, and in the United States, Japan and the countries of the European Union in particular, is to recognize that an isolated and purified form of a natural substance may be patented. For example, article 3.2 of the 1998 Directive on the Legal Protection of Biotechnological Inventions states that "biological material which is isolated from its natural environment or produced by means of a technical process" may be patentable. (Directive 98/44/EC of 6 July 1998, O.J. L213) Perhaps because this approach involves making a very fine distinction between discoveries and inventions, many developing countries have declined to follow it, instead choosing to deny patent protection to plant materials found in nature even if they have been isolated or purified by human intervention. According to one commentary, this patent ban is compatible with the TRIPs Agreement provided that it does not extend to plants with modified or artificial plant gene sequences, which often significantly vary from naturally occurring substances and thus are properly classified as inventions.

A second attribute of a patentable invention is that an applicant seeking such a patent must disclose the invention in a sufficiently clear and complete manner that permits a person skilled in the art to carry out the invention. (TRIPs, art. 29(1)). In addition to ensuring that the claimed invention in fact meets patent eligibility requirements, disclosure also permits third parties to access the invention for purposes of improving it or developing new inventions.

In the case of plant-related innovations, national laws permit inventors to satisfy the disclosure requirement either by a written description of the invention and/or by a deposit of the protected material (i.e. seeds, germplasm or other biological material). TRIPs does not require national governments to adopt any particular form of disclosure, nor does it specify the timing, manner or conditions of third party access. As a result, national laws vary widely on this point.

Novelty, inventive step and industrial application : These three cumulative eligibility requirements are mandated by article 27(1) of the TRIPs Agreement and are found in all national patent laws. Even among states with strong patent protection laws, however, the interpretation and application of each of these steps varies considerably.

Novelty : The principal objective of the novelty requirement is to ensure that the claimed invention cannot be found in the "state of the art" or "prior art" already in existence. These terms refer to the body of knowledge which is available to the public before the date of filing of an application for patent protection. Patent examiners establish both the novelty and inventive step requirements by comparing the claimed invention to the state of the art or prior art. A critical issue for plant-related innovations concerns the form in which prior art exists and the difficulty of accessing traditional knowledge as prior art.

Under United States patent law, for example, novelty can be negated if "the invention was patented or described in a printed publication in this or a foreign country or in public use or on sale in this country... " (35 U.S.C. Sec. 102(b)) The important point is description: foreign *uses* of claimed inventions are not considered as part of prior art. The Patent Cooperation Treaty (PCT) adopts a similar approach, limiting prior art to "everything which has been made available to the public anywhere in the world by means of written disclosure." (Rule 33.1 of Regulations under the PCT) Thus, in the United States and in countries following the PCT, even where foreign plant-related innovations are known, used or disclosed in other than written form, it may be possible for the inventor to obtain patent protection.

Article 54(2) of the European Patent Convention adopts a different approach, defining the state of the art to encompass everything made available to the public by means of written or oral descriptions, by use or indeed in any other way, prior to the filing of the application. Nevertheless, the absence of systemic databases concerning traditional knowledge, including knowledge relating to plants, has led examiners in various countries to grant patents for inventions that were later revoked after they were discovered to encompass prior art.

Inventive step : The inventive step requirement looks to the state of the art to determine whether the claimed invention is not obvious to a person skilled in the art. Although the majority of international treaties and national laws defined inventive step in this way, the application of the standard to plant-related innovations varies and may be a significant hurdle to patent protection given "the enormous speed of technological progress" in the field.

Industrial application : The industrial application requirement is concerned with an invention's practical utility, and in particular whether "it can be made or used in any kind of industry, including agriculture." (E.g. European Patent Convention, art. 7) This eligibility requirement does not appear to pose a significant obstacle to the patenting of new plant varieties, given their ready use in the fields of plant breeding, horticulture and agriculture.

Exclusions from Patentability

As noted above, article 27.3(b) of TRIPs permits WTO Members to deny all patent protection to "plants" and "plant varieties." Nevertheless, a state that chooses to provide such protection may nevertheless choose to exclude from patentability a narrower category of inventions, "the prevention within their territory of the commercial exploitation of which is necessary to protect *ordre public* or morality, including to protect... plant life or health or to avoid serious prejudice to the environment." (Art. 27(2)) Some commentators have asserted that this language grants states the discretion to exclude broad categories of inventions that further the erosion of genetic diversity on the theory that issues of public policy and morality include a state's sovereign right to control is natural resources.

However, because WTO dispute settlement panels have strictly interpreted clauses in the WTO Agreements that permit states to derogate from protected treaty rights, the more persuasive view is that exclusions on this basis must be fact-specific, narrowly drawn and no broader than required to achieve their purpose. (Moufang, 1998) Resolution of this debate in the context of plant-related innovations will not occur until the WTO dispute settlement bodies issue a ruling on the matter, an unlikely prospect so long as Members are free to exclude plants and plant varieties entirely from patentability.

Exclusive Rights

Once a patent is awarded, the owner of the patented product or process enjoys a broad panoply of exclusive rights. These include the right to prevent third parties from making the product, using the process or from using, offering for sale, selling or importing for those purposes the patented product or the product obtained by the patented process. (TRIPs, art. 28) States that choose to extend patent protection to plant-related innovations would thus be expected to protect products composed of plants and parts of plants and processes (including biological processes) for the production of plants. Although there is some dispute over whether traditional plant breeding methods (as opposed to creating varieties through genetic manipulation) can be protected by process patents, the practice of states that do extend patent protection to plants reveals that such methods have been protected.

Term of Protection

TRIPs article 33 imposes a minimum 20-year term of protection from the date the patent application is filed.

Exceptions and Limitations

As compared to limitations on plant breeders' rights permitted under the UPOV, the limitations on a patent owner's exclusive rights permitted under the TRIPs Agreement are far narrower. These limitations can be divided into exceptions to exclusive rights and compulsory licenses, which permit certain uses by third parties but require remuneration to the patent owner.

Exceptions : Article 30 of TRIPs permits WTO Members to adopt "limited exceptions" to exclusive patent rights provided that they do not "unreasonably conflict with a normal exploitation of the patent" and "do not unreasonably prejudice the legitimate interests of the patent owner, taking account of the legitimate interests of third parties." This three-part test is applicable to all patents protected by the TRIPs Agreement.

Canadian Patents Case : A WTO dispute settlement panel clarified the meaning of article 30 in Canada - Patent Protection of Pharmaceutical Products (Generic Medicines), WT/DS114/R (17 March 2000). The panel considered a complaint brought by the European Communities to two provisions of Canadian patent law. The first permitted uses of patented pharmaceuticals without the patentee's authorization for the purpose of obtaining approval of a generic product before the patent's term expired.

The second permitted production and stockpiling of the generic product for release immediately after the expiration of the patent. Both procedures were designed to permit the marketing of generic versions of pharmaceuticals promptly after the expiry of the patent. The panel decision confirmed that only the first exception is consistent with TRIPs. More importantly, the panel concluded that any exception that results in "a substantial curtailment" of the patent owner's exclusive rights is inconsistent with article 30. (WT/DS114/R, para. 7.36) If TRIPs were amended to require patent protection for new plant varieties, the standard adopted by the Canadian Patents panel would preclude Members from enacting many of the exceptions that are permitted under plant variety protection laws, such as the breeders' exemption and the farmers' privilege. The same prohibition on these privileges already exists in several countries that currently protect plant varieties and other plant-related innovations in their national patent laws.

Research and experimentation exception : Most national patent laws permit third parties to engage in experimentation or research related to the patented invention. Some states, however, including the United States, narrowly construe this exception to preclude unauthorized commercialization of any products or processes that result from such research if they encompass the patentable invention. In such countries, commercial activities permitted under the breeders' exemption found in plant variety protection laws are prohibited where the variety is protected by a patent.

Thus, for example, a breeder would not be permitted to cross patented seeds to produce improved varieties. Even in patent systems in which the experimental use exception tolerates some commercialization (Cornish, 1998), such as those found in Europe, it is likely that the commercial acts allowed under a traditional breeders' exemption would "conflict with a normal exploitation of the patent" and would thus amount "a substantial curtailment" of his or her exclusive rights in violation of TRIPs article 30. Thus, if patent protection of plant varieties were to become mandatory under TRIPs, a breeders' exemption such as that found in article 22,V of Mexico's patent law would be unlikely to survive a challenge before a WTO dispute settlement panel.

Farmers' privilege : A similar fate is likely to befall the exception permitting farmers to save and reuse seeds on their own land without the patent owner's permission, another exception traditionally allowed under plant variety protection laws. The United States does not recognize a farmers' privilege under its utility patent laws. Although such an exception has been adopted in article 11 of the European Union's 1998 Biotechology Directive (conditioned upon the payment of equitable remuneration except by small farmers), at least one commentator has questioned whether that exception is compatible with TRIPs article 30, because such an exception would unreasonably prejudice the "legitimate interests" of the patent owner.

Compulsory licences : TRIPs contains a complex set of rules that regulate when states may compel patent owners to license their products and processes to governments or to private parties. Although the TRIPs Agreement does not specify the grounds which justify the creation of compulsory licences, because article 5A(2) of the Paris Convention is incorporated by reference into article 2(1) of TRIPs, it can be inferred that such licences may be granted only to prevent "abuses which might result from the exercise of the [patent owner's] exclusive rights."

Even where such abuses exist, TRIPs article 31 imposes further conditions upon the granting of compulsory licences, including individual consideration of each case, prior negotiations with the patent owner seeking a voluntary licence, limitations on the scope and duration of compulsory licences and requiring their termination when the circumstances leading to their creation are no longer in effect. Most importantly, patent owners must receive "adequate remuneration," taking into account the value of the rights licensed. One area in which compulsory licences may affect plant breeders is that of dependent patents, which are defined as patents whose use requires the authorization of an earlier patent owner. Such patents are prevalent in plant breeding, where the creation of new varieties often occurs incrementally in the form of adaptations and improvements of existing varieties, as opposed to radically new innovations. Because incremental innovation often requires access to protected varieties, governments may seek to impose compulsory licences in favour of third party breeders who are unable to negotiate voluntary access to patented plant varieties. However, the compatibility of such licences with TRIPs is untested and governments seeking to impose them should ensure that they comply scrupulously with each of the many requirements set forth in article 31.

Because the patent compulsory licence provisions of TRIPs are far more detailed and narrow than the compulsory licences permitted under the 1978 and 1991 UPOV Acts, commentators are divided over their usefulness in the area of plant-related innovations. Some argue that such licences will be largely irrelevant to the field, whereas others claim that governments may adopt such licences "to ensure access to patented materials in order to attain specific agricultural objectives (e.g. availability of a given material for farmers) or food security...." Even if licences to achieve these objectives are permitted, however, the restrictions that TRIPs imposes significantly limit the ability of WTO Members to grant such licences in favour of third parties.

Sui Generis Protection for Plant Varieties under TRIPs

As noted above, TRIPs authorizes WTO Members to eschew patent protection for plants and plant varieties and adopt instead an "effective *sui generis* system" of protection. The following discussion identifies the core requirements of such a system, laying the groundwork for Part III which outlines the permissible options available to WTO Members seeking to grant IPRs to plant varieties while simultaneously seeking to achieve other objectives.

Identifying and Evaluating Competing Proposals for *Sui Generis* Systems

TRIPs does not define the term *sui generis* as it appears in article 27.3(b), nor does the treaty's drafting history shed any light on its meaning. As a general matter, the term *sui generis* is understood to mean "of its own kind" or "unique," a definition which does little to identify what sort of unique legal systems are permitted under the treaty. This lack of definitional clarity has allowed advocacy groups and commentators to advance different and often competing proposals for how effective *sui generis* protection should be structured, producing significant confusion for governments seeking to understand and implement their obligations under the TRIPs Agreement.

The discussion below identifies and evaluates the competing proposals. In brief, it concludes that if WTO Members choose to protect plant varieties using a *sui generis* right as opposed to a patent, they are required to create a distinct IPR applicable to plant varieties that complies with the core requirements and objectives of the TRIPs Agreement.

Sui generis systems that do not contain these core elements or conflict with them are not TRIPs-compatible and expose the WTO Members who adopt them to challenges under the WTO dispute settlement system. Nevertheless, the mandatory requirements of article 27.3(b) preserve significant leeway for national governments to work out the precise manner in which they will balance protection of IPRs against other international obligations and national objectives.

Sui Generis Systems as Defined by the UPOV Acts

Several commentators assert that a state adopting national legislation in compliance with either the 1978 UPOV Act or the 1991 UPOV Act has satisfied its obligations under article 27.3(b). Those advocating for compliance with the 1991 Act stress the more extensive IPR protections it provides to plant breeders, while commentators endorsing the 1978 Act as the appropriate *sui generis* standard stress that only the earlier Act was in force at the time TRIPs was being negotiated. In fact, however, the protection required by the two UPOV Acts is neither necessary nor sufficient for an "effective *sui generis* system." It is not necessary because article 27.3(b) does not require plant variety protection laws to contain the same subject matter, eligibility requirements, exclusive rights, terms of protection or other detailed provisions of either of the two UPOV Acts. And it is not sufficient because TRIPs requires WTO Members to structure their national IPR laws in ways that the two UPOV Acts do not.

Note, however, that developing counties that are parties to a "TRIPs plus" bilateral or regional trade or investment treaty may be required, by virtue of those treaties, either to join UPOV or to implement the standards set forth in one of the UPOV Acts. A country that has undertaken such obligations does *not* possess the same flexibility in designing a *sui generis* system to protect plant varieties as a country which has not undertaken such obligations.

For example, as explained in paragraphs 2.2.2.4 and 2.2.3.4 above, both the 1991 Act and the 1978 Act adopt the principle of national treatment, the same principle required by TRIPs. But under the UPOV Acts such treatment need be extended only to the nationals and residents of *other UPOV member states* and to legal persons having their headquarters in such states. Any WTO Member that limited national treatment in this way would unquestionably violate article 3 of TRIPs, which requires extension of national treatment to all other WTO Members whether or not they are also UPOV members.

However, to comply with article 3, a state whose plant variety protection law contains a UPOV-specific national treatment rule can easily amend its law to expand the number of states to which national treatment applies. Other UPOV-sanctioned rules may require more extensive changes.

For example, the 1978 Act allows a state to impose a reciprocity requirement in certain circumstances, limiting the right to apply for protection of a new variety to nationals or residents of those member states which also apply the Act to the genus or species to which that variety belongs. This reciprocal protection is incompatible with TRIPs' national treatment rule, and possibly its most favoured nation principle as well, both of which are inconsistent with a system of reciprocity.

That TRIPs' drafters did not intend either UPOV Act to be the exclusive model for *sui generis* protection of plant varieties is confirmed by their failure to refer to the Acts anywhere in the Agreement. By contrast, where the drafters intended Members to comply with standards found in preexisting international IPR treaties, they stated so expressly. Nevertheless, as explained in sections 3.3.1 and 3.3.2 below, most provisions of the two UPOV Acts are fully consistent with an effective *sui generis* system, and most countries have in fact adopted plant variety protection laws that are consistent with one or both UPOV Acts.

Core Requirements of an Effective Sui Generis System

According to a 1997 report prepared by two leading commentators, there are four core elements that any national plant variety protection law must contain in order to qualify as an "effective *sui generis* system" within the meaning of TRIPs article 27.3(b):(1) the law must apply to all plant varieties in all species and botanical genera; (2) it must grant plant breeders an IPR, i.e. the exclusive right to control particular acts with respect to those protected varieties, or at a minimum, the right to remuneration when third parties engage in certain acts; (3) it must provide national treatment and MFN treatment to breeders from other WTO Members; and (4) it must contain procedures that enable breeders to enforce the rights granted to them under such a law. The paragraphs that follow discuss each of these four requirements.

Protection of all plant varieties : Although both UPOV Acts require protection of only a specific number of plant varieties (paras. 2.2.1.1 and 2.2.3.1 above), a careful reading of the TRIPs Agreement reveals that WTO Members are required to protect all varieties. article 27.3(b) states that "Members shall provide for the protection of *plant varieties*" (emphasis added) without any further qualification or limitation. Inasmuch as article 27.3(b) is included in a provision of the TRIPs Agreement which carves out numerous and specific exceptions to patent protection, if the drafters had intended to allow states to limit the number or type of plant varieties subject to protection, they would have said so expressly. Indeed, the drafters were aware that both UPOV Acts permitted precisely such limitations, yet they chose neither to refer to those Acts nor to their limited scope of protection. As explained in paragraph 3.4.2.1 below, however, the obligation to protect all varieties does not require states to provide *the same level of protection* to each and every variety.

Protection of plant varieties with an IPR : If protection granted to plant varieties must take the form of an IPR, WTO Members would enjoy virtually unbounded discretion to choose the manner in which to protect plant varieties, for example through a system of taxation on seeds. The WTO's Appellate Body recently interpreted the text of the TRIPs Agreement, however, to require protection through an IPR. Nevertheless, because some commentators and NGOs have questioned whether plant variety protection is in fact a form of intellectual property included in the TRIPs Agreement, a brief discussion of this point is needed.

Article 1(2) of TRIPs defines "intellectual property" for purposes of the Agreement as "all categories of intellectual property that are the subject of Section 1 through 7 of Part II" of TRIPs. Protection for plant varieties appears in Section 5 relating to patents, and is thus a form of intellectual property protected by the Agreement. This interpretation is confirmed by the United States - Section 211 Omnibus Appropriations Act of 1998, AB-2001-7, WT/DS176/AB/R (2 January 2002). In that case, a WTO panel concluded that trade names were not a category of intellectual property protected by TRIPs because they were not expressly referred to in article 1(2) (an omission that also applies to plant varieties). The Appellate Body reversed this finding. As an example, it referred to Members' option to protect plant varieties "by *sui generis* rights (such as breeder's rights) instead of through patents," and concluded that *sui generis* rights were in fact a form of intellectual property protected by the treaty. (Id., para. 335)

The significance of this conclusion is that IPRs for plant varieties must take the form of other IPRs protected by the TRIPs Agreement. Specifically, governments must either grant to the owners of protected varieties (1) the right to exclude all third parties from engaging in certain activities with respect to those varieties (an exclusive rights approach) or, (2) at a minimum, the right to receive equitable remuneration when a third party engages in such activities.

Nothing in TRIPs requires states to adopt the exclusive rights approach, nor does it indicate which exclusive rights must be granted to new varieties. Nevertheless, the vast majority of IPRs are protected by exclusive rights, with compulsory licences used only in limited areas such as new industries or distribution networks, abuses of monopoly powers or where high transaction costs preclude private licensing mechanisms from developing. In addition, a number of economic studies have suggested that granting exclusive rights to the owners of intellectual property products is a more efficient mechanism for creating incentives to innovate than a compulsory licence approach, in which political factors or government inertia are likely to produce skewed incentives. (Merges, 1996)

National and MFN treatment applicable to rights granted : At least one respected commentator has argued that national and MFN treatment are inapplicable to a *sui generis* system protecting plant varieties. The theory advanced is that these two forms of treatment are inapplicable to IPRs not covered by the TRIPs Agreement and that plant variety protection is an IPR that is not covered by TRIPs. However, the text of TRIPs article 3(1), as interpreted by WTO dispute settlement jurists, indicates that plant variety protection must adhere to such obligations.

Thus, in the United States - Section 211 Omnibus Appropriations Act of 1998, AB-2001-7, 2 January 2002, the WTO Appellate Body stressed the critical importance of the national and MFN treatment rules in multilateral trade agreements (paras. 242 and 297), and it concluded that the obligation to grant such treatment applies to all subjects of intellectual property protected by TRIPs, a designation that includes *sui generis* protection of plant varieties. As a result, in the area of plant variety protection, each Member must grant "no less favourable treatment" to the nationals of all other WTO Members than it grants to its own nationals and it must also grant to such foreign nationals "any advantage, favour, privilege or immunity" provided to any other country. (TRIPs, arts. 3.1 and 4) Applied to plant varieties, this would mean that any IPR that a state grants to its own plant breeders must be granted to breeders from all other WTO Members, and that any IPR granted to breeders from one WTO Member must be granted to breeders from all WTO Members.

Enforcement of rights by private parties : The only qualification that article 27.3(b) imposes upon *sui generis* systems for protecting plant varieties is that they be "effective." Although TRIPs does not define this term, it refers to it in Part III dealing with enforcement of other IPRs against acts of infringement by third parties. For this reason, a WTO dispute settlement panel is unlikely to find a *sui generis* system to be effective unless it provides a meaningful opportunity for private parties to enforce their rights in protected varieties.

Specific enforcement measures are not mandated by the Agreement, however. Had the drafters intended greater specificity, they could have easily made a reference in article 27.3(b) to the detailed enforcement rules applicable to other IPRs in Part III of TRIPs. Although a *sui generis* system must be "effective," this does not mean that a state must devote a particular level of resources to enforcement measures. In particular, article 41.5 of TRIPs makes clear that the treaty does not create "any obligation with respect to the distribution of resources as between enforcement of intellectual property rights and the enforcement of law in general." As a result, what qualifies as "effective" in a state with abundant resources and well-developed systems of law enforcement will not be required of states with more limited resources or weaker law enforcement systems.

***Additional requirements of an "effective"* sui generis *system?* :** Beyond effective enforcement measures, it is uncertain whether a WTO dispute settlement panel would find that article 27.3(b) imposes additional requirements for a *sui generis* system to qualify as "effective." In general, international tribunals have concluded that rights granted in a treaty must be interpreted so as to make them effective rather than illusory.

Thus, if a WTO Member were to grant very limited and weak rights to breeders together with broad and extensive exemptions in favour of farmers or other users of plant germplasm, its plant variety protection law might not survive WTO scrutiny. Such a possibility is likely to arise only in extreme cases in which a WTO Member has enacted a plant variety protection system that is functional in theory but non-functional in practice.

States that implement the core TRIPs requirements in good faith - that is, states that grant breeders intellectual property rights and enforcement measures applicable to varieties in all species and botanical genera and that provide those same rights and measures to breeders from other WTO Members - are unlikely to have their laws challenged successfully.

Options Available to National Governments under Existing International IPR Agreements Protecting Plant Varieties and Plant Breeders' Rights

In order to identify the options available to national governments under international IPR agreements, this section begins by categorizing states according to the agreements to which they are parties. This classification scheme takes into account the fact that a state may be bound by more than one IPR-related treaty, a fact that may significantly limit its discretion. The section then identifies both the mandatory actions required of national governments in each category as well as the discretionary choices available to them. It also identifies trends in national laws which illustrate the manner in which states are complying with their particular set of international obligations.

Classification of States According to their International IPR Obligations

The discretion enjoyed by states to shape their plant variety protection laws to balance the protection of IPRs against other societal concerns is dependent upon the international agreement or agreements to which they are parties. The following table identifies that discretion at a broad level of generality. It does so by listing the mandatory components of plant variety protection that a state must adopt depending upon whether it is a member of (1) WTO only; (2) WTO and the 1991 UPOV Act; (3) WTO and the 1978 UPOV Act; and (4) either the 1991 UPOV Act or the 1978 UPOV Act only; or (5) no IPR agreement relating to the protection of plant varieties.

Note that some developing states, by virtue of having ratified a "TRIPs plus" bilateral or regional trade or investment treaty, may be required to adhere to some or all of the provisions contained in either the 1991 UPOV Act or the 1978 UPOV Act, *even if they are not members of UPOV itself.* Developing states that are parties to such "TRIPs plus" treaties should carefully consult the text of these agreements to determine the precise scope of their supplementary IPR obligations, including for purposes of applying the classification scheme set forth below.

For example, a developing country that is a WTO Member but not a member of UPOV may nevertheless be required to adhere to all of the provisions of the 1991 UPOV Act by virtue of having ratified a "TRIPs plus" treaty. In such a case, the state would be obligated to implement the mandatory components of plant variety protection contained not only in TRIPs but also in the 1991 UPOV Act, and its discretion to balance the protection of IPRs against other competing policy objectives would be reduced accordingly.

Cumulative Treaty Obligations

Thus, for example, a state that is a party to a UPOV Act and also to the TRIPs Agreement must comply with its obligations under both treaties. Given that each treaty contains somewhat different but mostly cumulative and consistent rules concerning the protection of plant varieties, a state that is a party to both treaties will necessary have more extensive obligations to protect plant varieties and thus will enjoy concomitantly less discretion.

Conflicting Treaty Obligations

Where the provisions of two treaties are in direct conflict, the rule is far less settled. If two agreements relate to the same subject matter and the states parties to both agreements are the same, article 30 of the Vienna Convention on the Law of Treaties specifies that the agreement that is later in time is given effect. The obligations in the TRIPs Agreement are thus likely to prevail over any conflicting obligations in the 1978 UPOV Act, such as the ban on protecting varieties within the same genus or species with both a breeders' right and a patent. (1978 Act, art. 2(1)) Because the 1991 Act entered into force in 1998 and TRIPs entered into force in 1995 it may be argued that the 1991 UPOV Act is the later-in-time agreement.

However, it is unlikely that any conflicts issue will arise between the two treaties, inasmuch as nothing in TRIPs requires conduct that the 1991 UPOV Act forbids. (Such a conflict might arise, however, in the unlikely event that TRIPs were interpreted to impose more restrictive exclusions to breeders' exclusive rights than the mandatory exclusions found in article 15 of the 1991 UPOV Act.) It can therefore be argued that the two treaty systems are fully compatible, with TRIPs merely augmenting the plant variety protection requirements of the UPOV Acts.

Identifying the Level of Discretion Available to States in each Treaty Classification

In general, the discretion a stateenjoys in limiting the scope of intellectual property protection provided to plant varieties increases.

WTO and UPOV 1991 Act Members

States that are members of both TRIPs and the 1991 UPOV Act have the least discretion. This designation currently applies to 27 nations: Australia, Austria, Belarus, Bulgaria, Croatia, the Czech Republic, Denmark, Estonia, Finland, Germany, Hungary, Israel, Japan, Kyrgyzstan, Latvia, Lithuania, the Netherlands, Moldova, Poland, Romania, Singapore, Slovenia, South Korea, Sweden, Tunisia, the United Kingdom and the United States of America.

Mandatory Requirements

These 27 states must extend protection to all plant varieties, comply with TRIPs' national MFN treatment obligations and adopt effective enforcement measures. The adoption of these rights - (1) production or reproduction, (2) conditioning for the purpose of propagation, (3) offering for sale, (4) selling or marketing, (5) exporting, (6) importing and (7) stocking for any of these purposes - fully satisfies their obligation under TRIPs to protect varieties with an IPR. Finally, states that have agreed to these dual obligations must also comply with all of the other provisions of the 1991 Act, including its terms of protection and limitations on the breeders' exemption and farmers' privilege.

Options and Implementation Issues

A state whose plant variety protection laws are already in compliance with the 1991 Act needs to take only limited steps to comply with article 27.3(b) of TRIPs, including immediate protection of all varieties and extension of national treatment to all WTO Members. The most difficult hurdle is likely to involve enacting effective mechanisms for plant breeders to enforce their rights. However, far more detailed enforcement measures must be adopted for the other forms of intellectual property protected by TRIPs. A state may therefore choose to extend to breeders the same domestic enforcement opportunities that it is required to provide to the owners of other IPRs.

National Implementation

In information submitted to the TRIPs Council in May 1999, the following states indicated that they had adopted plant variety protection laws modelled on the 1991 Act: Australia, Bulgaria, Members of the European Communities, Japan, Morocco, Poland, Romania, South Korea, South Africa, Switzerland and the United States of America. (Other states parties to both UPOV 1991 and TRIPs did not submit information to the Council.) Of the states adopting such laws, Morocco, South Africa and Switzerland had not ratified the 1991 Act.

Bilateral IPR Agreements and UPOV Membership

A separate issue concerns whether WTO Members will continue to join UPOV after the entry into force of the TRIPs Agreement. Some commentators predicted that UPOV membership would stagnate as states capitalized on the greater discretion provided to them under the *sui generis* option of article 27.3(b). Since TRIPs entered into force in 1995, however, 28 nations have become parties to the 1978 or 1991 UPOV Acts. (States Parties to the UPOV, Status on 30 June 2004) The continued expansion of UPOV membership can be traced in part to the recent proliferation of "TRIPs plus" treaties.

WTO and UPOV 1978 Act Members

This esignation currently applies to 24 nations: Argentina, Bolivia, Brazil, Canada, Chile, China, Colombia, Ecuador, France, Ireland, Italy, Kenya, Mexico, New Zealand, Nicaragua, Norway, Panama, Paraguay, Portugal, Slovakia, South Africa, Switzerland, Trinidad and Tobago and Uruguay. States in this category enjoy somewhat greater discretion as a result of the more limited protection of plant breeders' rights contained in this earlier UPOV Act.

Mandatory Requirements

States parties to both agreements must extend protection to all plant varieties, comply with TRIPs' national and MFN treatment rules and adopt effective enforcement measures. They must also comply with all of the other 1978 Act requirements, including its eligibility requirements, terms of protection, exclusive rights and mandatory breeders' exemption. As compared to the 1991 Act, however, breeders' exclusive rights are more limited, terms of protection for varieties are shorter and exceptions and limitations are broader.

Options and Implementation Issues

States that become Members of the WTO after joining the 1978 UPOV Act and adopting laws to comply with that Act face a similar situation to states that are parties to both the WTO and the 1991 Act. To fully comply with TRIPs, these 1978 Act member states must modify their national laws to protect the four core requirements of article 27.3(b) and they must remove all provisions of their laws which impose a reciprocity requirement as a condition for protecting varieties of foreign breeders. In addition, states in this category may choose to modify their laws to incorporate some or all of the standards found in the 1991 Act without actually becoming a member of that Act. Their refusal to do so, however, does not violate article 27.3(b), inasmuch as the standards found in the 1978 Act satisfy their obligation to protect plant varieties with a *sui generis* IPR.

Trends in National Laws

In information submitted to the TRIPs Council in May 1999, Canada, the Czech Republic, Hungary, Norway and New Zealand (the only states submitting information) indicated that they had adopted plant variety protection laws that complied with the 1978 Act. At the time of these submissions, the Czech Republic and Hungary were members of the 1978 Act. They have since acceded to the 1991 Act.

1991 or 1978 UPOV Act Members Only

There are few states outside of the WTO that protect plant varieties solely by virtue of being UPOV members. And both are observer governments with the WTO. These nations face a somewhat different set of international legal obligations.

Mandatory Requirements

They must comply with all of the numerous requirements of the UPOV Act to which they are a party. However, they have no obligation to provide effective enforcement measures to breeders, their national treatment obligations are limited to those states that are members of the same UPOV Act and, in the case of those states bound by the 1978 Act, they need not protect all plant varieties and may impose certain reciprocity requirements on those varieties they do protect.

Options and Implementation Issues

Provided that their laws are in compliance with the relevant UPOV Act and unless they have ratified other international IPR agreements, these states need make no additional changes to their national plant variety protection laws.

Member of WTO Only

States that are parties to the TRIPs Agreement but are not members of either UPOV Act enjoy considerably greater discretion than those that are members of either UPOV Act alone. By virtue of their adherence to TRIPs, these states are required only to comply with the four core obligations of article 27.3(b). However, those countries that have ratified “TRIPs plus” treaties that require them to comply with one of the UPOV Acts). Given the large number of states in this category (and the additional number of states likely to enter this category if their efforts to join the WTO succeed), a more detailed discussion of the numerous options available to them in balancing protection of plant breeders' rights against the competing policy objectives is provided in section 3.4 below.

Not a Member of WTO, UPOV or other IPR Agreements

Nations in this category have no obligation to protect plant varieties or breeders' rights in any form. Nevertheless, the mandatory requirements and discretionary choices facing other governments are likely to become matters of significant concern to these states as the world trading and intellectual property protection systems expand. Attention to IPR issues will be of particular importance when these states implement the obligations of other international treaties to which they may be parties, such as the biodiversity goals of the CBD or the access requirements of the ITPGR. States in this category should consider whether it is sound policy to enact laws that are incompatible with TRIPs and the UPOV, since such an action will preclude their membership in these treaty systems unless they undertake a further revision of their national laws.

In addition, unilateral actions by or bilateral agreements with industrialized nations may pressure these states to comply with international IPR standards even if they remain outside of the WTO or UPOV. For these reasons, states that are not yet Members of the WTO should give careful attention to the discretionary choices available to WTO Members analysed in the next section.

2

Genetics Embryology

Embryology

Embryology is the branch of developmental biology that studies the beginning and early growth of sexually reproducing organisms. This life science speciality focuses primarily on *embryogenesis* (the formation and development of embryos). In humans, the term embryo is traditionally reserved for the first two months of development. After that point, the term "embryo" is replaced by the term "fetus," which then applies until birth. At the beginning of the individual's development, the entity is a single cell. After two months, it has limbs, distinct fingers and toes, internal development, and countless cells. So the term "*embryo*" applies to an individual throughout a vast range of developmental change.

Since the time Charles Darwin published *The Origin of Species*, embryology has been used to support common descent. In fact, Darwin referred to embryological homology as the *strongest single class of facts* that existed to support his theories. A theory later put forth by Ernst Haeckel, known as the biogenetic law, asserted that the evolutionary history of an organism was recapped during embryo development.

Although the biogenetic law is now discredited, in recent years, embryology has reemerged as a tool used by evolutionary biologists that attempt to establish phylogenetic relationships by identifying developmental similarities between taxonomic groups. *Evolutionary Developmental Biology* is a merging of developmental biology and evolutionary biology that is commonly known as "*evo-devo*".

Embryogenesis

Embryogenesis begins when a sperm fertilizes an egg and creates a single cell that has the potential to form an entire organism. In the first hours after fertilization, this cell divides into identical cells. Approximately 4 days after fertilization and after several cycles of cell division, these cells begin to specialize, forming a hollow sphere of cells, called a blastocyst. The blastocyst has an outer layer of cells, and inside this hollow sphere, there is a cluster of cells called the inner cell mass. The cells of the inner cell mass will go on to form virtually all of the tissues of the human body.

Although the cells of the inner cell mass can form virtually every type of cell found in the human body, they cannot form an organism. Therefore, these cells are referred to as pluripotent, that is, they can give rise to many types of cells but not a whole organism. Pluripotent stem cells undergo further specialization into stem cells that are committed to give rise to cells that have a particular function. Examples include blood stem cells that give rise to red blood cells, white blood cells, and platelets, and skin stem cells that give rise to the various types of skin cells. These more specialized stem cells are called multipotent—capable of giving rise to several kinds of cells, tissues, or structures.

Cleavage

After fertilization, the zygote proceeds immediately to the first cleavage (cell division) and subsequent cell divisions follow rapidly. The zygote is a very large cell, but the first waves of rapid cell division occur without increase in cell volume. The result is a closely bound mass of cells each of more typical cell size. At this stage the cells are called blastomeres, and the organism as a whole is called a morula from the time it has 16 blastomeres to the next stage. As it is undergoing this very rapid cell division, the organism is also migrating down the uterine tube toward the uterus.

Vertebrates, such as humans, have bilateral symmetry and therefore polarity in three dimensions (head-tail, or back-front, and left-right). Establishing polarity is one of the most basic manifestations of emerging specialization. But the egg is roughly spherical, and it is not readily apparent how polarity is established. Although it had been shown long ago that the point of sperm entry determines the plane of first cleavage (and thus subsequent ones) in amphibian eggs, mammals were believed until recently to remain spherically symmetrical until later in development. Recent data on mammalian zygotes, however, suggests that the point of sperm entry may similarly determine the cleavage plane. Even the first two cells resulting from the first cleavage may have different propensities, which persist through the next divisions as the progeny of one cell tend to become the body of the offspring and progeny of the other cell become the embryo's contribution to the placenta and other supporting structures. The word "fate," however, might be too strong, because the cells of such very early embryos are resilient to perturbations—if one cell is removed, the remaining ones can compensate.

Gastrulation

Gastrulation begins when fluid starts to accumulate between the blastomeres of the morula. The fluid-filled spaces run together, forming a relatively large fluid-filled cavity. At the point when the cavity becomes recognizable, the organism is called a blastocyst. The outer cells of the blastocyst, especially those around the blastocyst cavity, assume a flattened shape. The flattened cells of the exterior blastocyst are the trophoblast.

They become the embryo's contribution to the placenta and other supporting structures. On one side of the blastocyst is a group of cells that project inside into the blastocyst cavity; this is the inner cell mass, or embryoblast, and its progeny form the body of the new offspring.

The cells of the inner cell mass can give rise to progeny differentiating into all the types of cells in the adult body, so they are called pluripotent. They have not usually been described as totipotent because, the inner cell mass having already differentiated from trophoblast, the cells of the inner cell mass were believed to be no longer able to give rise to the cells of the trophoblast.

Recent work, however, describes culture conditions under which human embryonic stem cells can differentiate to trophoblast cells. Although the new offspring itself develops only from the inner call mass, the trophoblast is not just passive padding. Its progeny are the essential and specialized connection between the embryonic and maternal systems. Embryonic stem cells can be isolated from the inner cell mass.

Organogenesis

The basic structures and relations of all the major organ systems of the body emerge during the fourth through the eighth weeks of embryonic development. First, the embryo folds in several ways so that the flat linear structure distinguished by neural tube flanked by somites become roughly C-shaped. The effect of this is to bring the regions of the brain, gut, and other internal organs into their familiar anatomical relations. During the fourth week the neural pores, the ends of the neural tube "zipper," close.

First the one at the cranial or head end, which is called the anterior or rostral pore, closes, and later the caudal or tail-ward pore closes. Closure of the neural pores completes the closure of what will become the central nervous system. Also during the fourth week, limb buds become visible, first buds for arms and later for legs. Further, two accumulations of cells along the neural tube become distinguishable: the alar plate and the basal plate. Cells of the alar plate go on to become mostly sensory neurons, while basal plate cells give rise mostly to motor neurons. Already while the neural tube is closing, its walls along the cranial area are thickening to form early brain structure. Cranial nerves, for example the nerves for the eye and for the muscles of the face and jaw, also are beginning to develop at this time. The embryonic brain develops rapidly in both size and structure especially during the fifth week, and the optic cup that will form the retina of the eye becomes visible as well.

Support for Darwinism

Von Baer's Law

Several decades before Darwin published the *The Origin of Species*, a German embryologist Karl Ernst von Baer had shown that the embryos of some vertebrates looked very similar during particular stages of development. This observation becme known as *von Baer's law*. After reading a summary of von Baer's law in 1842, Darwin concluded that embryonic resemblances were a very strong argument in favour of common descent. In The Origin of Species (1859) Darwin stated *"Community of embryonic structure reveals community of descent"*, and in a 1860 letter to Asa Gray stated that *"embryology is to me by far the strongest single class of facts in favour of change of forms"*. Despite Darwin's use of embryology to support his views, Von Baer rejected the theory of common descent and did not regard embryological similarities as evidence for evolution.

Recapitulation Theory

The theory of recapitulation was a subsequent attempt to apply embryological similarity in support of Darwinism, which was first put forth in 1866 by Ernst Haeckel. He coined the phrase *ontogeny recapitulates phylogeny* to summarize the view.

- Ontogeny is the development of the organism.
- Phylogeny is the evolutionary history.

Haeckel suggested that *gill slits* and a *tail* could be seen during embryo development in organisms that did not possess them as adults. It was asserted that these organism must have possessed structures during their evolutionary history that were lost in adult stages due to natural selection. His theory held that during embryo development an organism progresses through stages in its evolutionary history-a theory that became known as the *Biogenetic law*.

It is important to note that the *gill slits* are not slits at all, but merely the folds in the region of the embryo's throat that are never involved with gas exchange (O_2/CO_2), which is accomplished by the umbilical cord. These folds, now called *pharyngeal arches*, later develop into the outer and middle ear as well as the bones, muscles, nerves, and glands of the neck. The appearance of an embryonic *tail* is simply due to the rapid growth of the spinal cord causing it to become longer than the body, and forcing the body to curl. This curling contributes to the folds visible in the neck region, in the same way a person develops a double chin when bending the neck forward.

Haeckel's Fraud

In an attempt to prove his theory true, Ernst Haeckel committed one of the most infamous examples of scientific fraud. In 1868, he published the series of embryos in the picture at right showing various living things placed side by side in the attempt to give the impression that there was significant similarity among them.

The embryos shown in the earliest stages were altered to make them look more similar than they are in actuality. Embryologist Jonathan Wells notes that Haeckel was also very selective about the embryos he chose for his illustration. The four right-hand columns in the diagram at right are all from the same taxonomic subclass *Eutheria* (Placental mammals). Marsupials and monotremes were excluded, as were cartilaginous fish (sharks) and others.

The fraud was exposed only months following the publication of these engravings by L. Rütimeyer, professor of zoology and comparative anatomy at the University of Basel, and corroborated by other contemporaries such as William His Sr, professor of anatomy at the University of Leipzig, who published their own comparisons showing significant differences (picture at left).

In 1997, the full extent of Haeckel's fraud was brought to light when Michael K. Richardson, a lecturer and embryologist at St George's Hospital Medical School, London, published photos of actual embryos illustrating the discrepancies between Haeckel's popular diagram and genuine embryos. Richardson described the situation in an interview in *Science* magazine as follows: *It looks like it's turning out to be one of the most famous fakes in biology*.

The effort involved an international team of embryologists who examined and photographed *"the external form of embryos from a wide range of vertebrate species, at a stage comparable to that depicted by Haeckel"* because up until that time it was apparent that *"no one has cited any comparative data in support of the idea"*.

They analysed 39 different animals including those published by Haeckel and many that were not, such as marsupials, tree-frogs, snakes, and alligators, and they found significance differences with little conservation at the stage of development Haeckel claimed was most similar. In fact, the species in the fraudulent publication are so different that it is assumed the drawings made by Haeckel could not possibly have been done from real specimens.

Richardson summarizes the extent of their findings:

> *"We find that embryos at the tailbud stage – thought to correspond to a conserved stage – show variations in form due to allometry, heterochrony, and differences in body plan and somite number. These variations foreshadow important differences in adult body form. Contrary to recent claims that all vertebrate embryos pass through a stage when they are the same size, we find a greater than 10-fold variation in greatest length at the tailbud stage. Our survey seriously undermines the credibility of Haeckel's drawings, which depict not a conserved stage for vertebrates, but a stylised amniote embryo."*

In an interview for the London Times, Richardson is quoted as saying:

> *"This is one of the worst cases of scientific fraud. It's shocking to find that somebody one thought was a great scientist was deliberately misleading.... What he did was to take a human embryo and copy it, pretending that the salamander and the pig and all the others looked the same at the same stage of development. They don't ... These are fakes."*

Textbook Use

Despite the early exposure of this fraud by his contemporaries and following the almost universal recognition by the scientific community, Haeckel's embryo comparisons continue to appeared in textbooks in support of common descent. Examples of recent use include the 2004 edition of Biology: The Unity and Diversity of Life by Starr and Taggart., and the 1992 edition of Biology, a college level introductory biology text that states: *"In many cases the evolutionary history of an organism can be seen to unfold during its development, with the embryo exhibiting characteristics of the embryos of its ancestors. For example, early in their development, human embryos possess gill slits like a fish..."*

In 2000, the eminent evolutionist Stephen Jay Gould calls the continued use of these embryos in textbooks *atrocious*:

> *"We do, I think, have the right to be both astonished and ashamed by the century of mindless recycling that has led to the persistence of these drawings in a large number, if not a majority, of modern textbooks. "*

Homology

Similarities that exist between different organisms are cited by evolutionists as one of the most compelling lines of evidence for their theory. *Homologous* features are shared design concepts that are generally assumed by evolutionists to be the result of a shared ancestry and therefore offering support for the

Darwinian theory of common descent. In The Origin of Species, Darwin referred to the *similar bones in the same relative positions* in the hand of a man and several other mammals as *specifically what the theory of common descent would expect*. The 2008 edition of Prentice Hall Biology by Miller and Levine illustrates that this concept is commonly held wherein it states: *In their early stages of development, chickens, turtles, and rats look similar, providing evidence that they shared a common ancestry*.

Much like gene comparisons, the similarities that are evident during embryo development (*von Baer's law*) are assumed outright to be due to phylogenetic relationships and therefore homologous. Furthermore, embryo homology is considered to be such powerful evidence of common ancestry that patterns during embryogenesis are used to determine if shared morphological structures are truly homologous (due to shared ancestry) or merely analogous (due to convergent evolution).

Miller and Levine define homology in such terms: *structures that have different mature forms in different organisms but develop from the same embryonic tissues*. The reemerging of the early 19th century *"Von Baer's law"* has stimulated a new biological discipline called Evolutionary Developmental Biology, which is a collaboration of Evolutionary Biology and Developmental Biology, popularly known as Evo-devo.

This area of scientific inquiry draw from the beliefs of Darwinism that all organisms have descended from the same ancestor, and generally assumes that embryonic developmental mechanisms should be shared (conserved). Drawing from these assumptions, evolutionary developmental biologists compare the developmental processes of different animals and plants in an attempt to determine the ancestral relationship between organisms.

Developmental biologist William Ballard a well known elder developmental biologist at Dartmouth College comments on the impact of *von Baer's law* on modern embryology.

"Thus, the energy of investigators and particularly students is diverted into the essentially fruitless 19th century activity of bending the facts of nature to support second-rate generalities of no predictive value. Though enthusiasm for Haeckel's (1900) recapitulation `law' died out, unfortunately the popularity of Von Baer's `laws' of 1828 was renewed. In order to defend the latter's descriptive statements that general characters appear before special characters as an egg develops and that the less general and finally the specific characters trail along later, we have to decide intuitively that certain characters are of `morphological significance' and others are not. When referring to vertebrates, we have to use words like blastula and gastrula in such a way as to imply that things that are vastly different from each other are really very much the same. "

Differences among organisms have traditionally been assumed to become more significant as development proceeds, and therefore earlier embryonic stages should resemble one another more than latter stages. This view can be found presented in popular biology textbooks. Although these assumptions have persisted into modern times, it has been recognized by embryologists since Darwin that this doctrine of developmental biology *"is not in accordance with the facts of development"*, according to British embryologist Adam Sedgwick (1894) who also stated: *there is no stage of development in which the unaided eye would fail to distinguish between them with ease"*. Modern embryologists confirm this as well, and have further challenged the erroneous nature of these assumptions. Recent work has revealed that embryos are not more similar during the earlier stages of cleavage and gastrulation, but in fact, are most alike at the later stage illustrated by Haeckel and recently photographed by Richardson.

In the gastrulation stage, fish show dramatic differences when compared to amphibians, and neither are similar to reptiles, birds, or mammals. Organisms begin life very dissimilar, then become somewhat alike during embryo development, then diverge again as adults. Evo-devos (evolutionary developmental biologists) refer to this pattern as the *developmental hourglass*. William Ballard adds that *only by semantic tricks and subjective selection of evidence can we claim that "gastrulas" of shark, salmon, frog, and bird are more alike than their adults.*

Ballard continues:

> *"In fact, the most obvious structural characteristics of either the eggs or the cleavage stages of a shark, a salmon, a frog, a bird, or a mammal are unique each to its own class, not generally shared. We would not consider them very much alike unless we had been taught so at a very early age. Very few vertebrates pass through a stage which can strictly be called a blastula. The embryo in its period of most active morphogenetic movements is usually called a gastrula, but as all agree this word has no morphologic meaning anymore. Each class of vertebrates (in mammals we might almost say each particular order) develops and then loses its own set of temporary structures-like the parade ground `formations of maneuver' during this period. "*

Another significant challenge for evo-devo is the recognition that organs and structures develop from different groups of embryo cells in organisms that are believed to share a common ancestry. Fundamental structures, like the alimentary canal, kidney, or forelimb are commonly found to form from different embryological tissues. The alimentary canal (digestive tract) is formed from the roof of the embryonic gut cavity in sharks; whereas in frogs it is formed from the gut roof and floor; and in birds and reptiles it is derived from the lower layer of the embryonic disc or blastoderm. Vertebrate forelimbs, which were used as an example of homology by Darwin and can be found in hundreds of textbooks as proof for evolution, also develop from different body segments in different species. Rather than being proof of Darwinism, recent embryological studies have shown that forelimbs are not developmentally homologous. The forelimbs in the newt develop from embryo trunk segments 2 through 5; in the lizard they develop from segments 6 to 9; in humans they develop from segments 13 through 18. The development of the vertebrate kidney also conflicts with the view of homology. Evolutionist Michael Denton notes these conflicts in Evolution: A Theory in Crisis.

> *"In fish and amphibia the kidney is derived directly from an embryonic organ known as the mesonephros, while in reptiles and mammals the mesonephros degenerates towards the end of embryonic life and plays no role in the formation of the adult kidney, which is formed instead from a discrete spherical mass of mesodermal tissue, the metanephros, which develops quite independently from the mesonephros.*

Studies of the genes governing development have added even further difficulties for evo-devo. According to the views of neodarwinists, homologous features are produced by genes that have been passed down by common ancestors. Because similar genes are unlikely to have originated independently through random mutation, sequence similarity between organisms indicates they share an ancestral relationship (they are homologous). Features produced by similar sequences are then also inferred to be phylogenetically homologous.

For decades it has been recognized that homologous organs were frequently produced by different genes, but more recently, evo-devo biologists discovered that certain developmental genes were highly conserved (similar) in radically different organisms. These homeotic genes are short sequences that act as regulatory switches affecting the development of structures such as eyes and legs. When mutated, homeotic genes can transform one body part into another, such as an antenna into a leg. In organisms as vastly different as mammals and fruit flies, these genes are so similar that the mouse gene can regulate the development of the compound eye of the fruit fly. In such examples, the conflict with Darwinian assumptions lies in the fact that non-homologous structures are produced (or at least regulated) by supposedly homologous genes. As Jonathan Wells states very poignantly: *homology, whether defined morphologically or phylogenetically, cannot be attributed to similar developmental pathways any more than it can be attributed to similar genes.*

The famous embryologist Sir Gavin de Beer challenged his colleagues with the following question which remains unanswered to this day.

"But if it is true that through the genetic code, genes code for enzymes that synthesize proteins which are responsible (in a manner still unknown in embryology) for the differentiation of the various parts in their normal manner, what mechanism can it be that results in the production of homologous organs, the same "patterns", in spite of their not being controlled by the same genes? I asked this question in 1938, and it has not been answered.

Worldviews

While Darwin was convinced that embryological homology was the *strongest single class of facts* to support his views. Modern embryology has instead challenged the Darwinian theory of common descent by illustrating that seemingly homologous structures, such as the mammalian forelimb, developed along different pathways, and are therefore not homologous at all.

While the Darwinists see these similarities as evidence of common descent, the creationist belief system causes us to generally conclude that shared structures, processes, or genes are merely reflections that they were designed by the same creator. Rather than being evidence of shared ancestry, homologies are examples of brilliant and well functioning designs that have been applied to multiple organisms, much in the same way as human designers apply concepts.

The Rise of Fish Embryology in the Nineteenth Century

It is particularly appropriate to write about the history of fish embryology at this time. History is often said to move in cycles, and this seems to be true of the study of fish development. After a period of prominence during the nineteenth and early twentieth centuries, the study of fish development, except in the hands of a stalwart few such as Oppenheimer and Trinkaus working with Fundulus and Ballard and Devillers working with the trout and other fishes, was overshadowed by developmental studies of other organisms, such as amphibians, the domestic fowl, sea urchins, and assorted invertebrates. For obvious reasons, the study of fish development continued to be pursued by fisheries biologists. With the popularization of the zebrafish Danio rerio as a "model of vertebrate development" and a revival of interest in the relationship between develoment and evolution, the cycle has come full turn and the study of fish development again has a place in the sun.

My objective is to present an outline of the historical development of the science of fish embryology in the nineteenth century, broadly defined as 1789-1914. My plan is to assess the state of knowledge at the outset of the period, to document the advancement of descriptive embryology of teleosts and elasmobranchs, to consider the rise of comparative embryology of fishes associated with evolutionary studies, and to search out the beginning of experimental, physiological, and biochemical analyses of fish development.

When discussing the development of fishes, it is prudent to recall that fishes are the most diverse group of vertebrates. They should not be treated as a distinct, relatively homogeneous taxonomic unit. Rather, the term "fishes" applies to a grade of ectothermic, aquatic craniates and vertebrates and not to a distinct taxonomic group (Atz, 1985; Bruton, 1990; Nelson, 1994). Extant fishes are divided into five classes, namely Myxini (hagfish), Cephalaspidomorphi (lampreys), Chondrichthyes (cartilaginous fishes), Sarcopterygii, and Actinopteryg. Chondrichthyes contains two taxa, namely the subclass Elasmobranch (living sharks, rays, and skates) and the Holocephal (ratfishes or chimaeras) that diverged at a relatively early time, i.e., the Devonian-Carboniferous boundary.

The class Sarcopterygii (flesh-finned fishes) contains the Actinistia (the living coelacanth), the Dipnoi (lungfishes) and the tetrapod vertebrates. The class Actinopteryg (ray-finned fishes) comprises five groups, namely Cladistia (bichir and reedfish), Chondrostei (sturgeons and paddlefish), Ginglymodi (gars), Halecomorphi (Amia calva, the bowfin), and the Teleostei (teleost fishes). Thus, the five classes of fishes contain eleven major taxonomic groups with an estimated total of 25,000 species. Each class is as distinct from the others as it is from the four other classes of vertebrates. The theme of eleven major taxonomic groups of fishes that is used in the ensuing presentation to emphasize the broad phylogenetic relationships of fishes and other vertebrates, tends, however, to obscure the extraordinary phylogenetic diversity within the teleosts. Teleostei, a monophyletic group, contains 24,000 species in 38 orders and 126 families compared to the 1,000 species, 19 orders, and 56 families in the ten other groups of fishes and the 25,000 species of tetrapods.

Not only are fishes the most diverse group of craniates, but they also display the greatest diversity of life history styles. Most fishes are oviparous, but viviparity is estimated to have independently evolved 42 times in five of the eleven major groups of fishes, all within the Chondrichthyes and Osteichythes. The phylogenetic diversity of fishes is reflected in the diversity and variability of their developmental patterns. Craniate development based on cleavage and subsequent patterns of gastrulation and embryogenesis is holoblastic, meroblastic, or transitional between the two. In holoblastic eggs, cleavage of the ooplasm is complete by the blastula stage. The morphogenetic movements of gastrulation occur in the three-dimensional space of the entire egg and involve most of the embryonic mass.

They often lead to the occlusion of the blastocoele. Neurulation is synchronous in time and space. Holoblastic development is considered the less specialized and more primitive craniate developmental pattern because it is characteristic of most echinoderms and non-craniate chordates as well as the more primitive representatives of various craniate lineages, namely lampreys, sturgeons, bichirs, lungfish, and most amphibians. In the meroblastic egg, most of the yolk mass remains uncleaved at the blastula stage. Morphogenetic movements are confined to a portion of the egg volume, i.e., the blastoderm, and involve only a portion of the embryonic mass.

Gastrulation and neurulation tend to be spatially and temporally asynchronous. Evolution of the meroblastic pattern appears to have involved qualitative and quantitative changes in the ooplasm and yolk components of the egg, namely an increase in total yolk content, an increase in the amount of yolk relative to ooplasm, a change in the type of yolk, and a change in the relative disposition of ooplasm and yolk.

Not only did the changes in yolk-cytoplasm relationships of the egg affect cleavage patterns, but they also imposed a set of physical or physiological constraints on the morphogenetic movements of embryo formation, the so-called holoblastic-meroblastic barrier. Consequently, new or highly modified patterns of gastrulation and embryogenesis co-evolved with the evolution of the meroblastic egg.

The meroblastic pattern of development appears to have independently evolved in six craniate lineages, namely hagfish, sharks and chimaeras, teleost fishes, the living coelacanth (?), one group of amphibians (the caecilians) and amniotes. In each instance, different morphological patterns of gastrulation and embryogenesis have evolved.

Thus, although meroblastic cleavage is superficially similar in hagfish, elasmobranchs, and teleosts, the morphological events of embryo formation differ considerably. These points and others are summarized in part by Collazo et al. (1994) who present a phylogenetic perspective and an evolutionary scenario for teleost gastrulation. Paradoxically, whether development proceeds according to a holoblastic or meroblastic pattern, the end product is a stereotypic craniate embryo. Recent advances in the study of the molecular basis of embryogenesis should help resolve this paradox.

Descriptive Embryology

Beginning with Aristotle, various authors have summarized the existing knowledge of fish development. In most instances, they did so in conjunction with larger works that dealt with the natural history or systematics of fishes, e.g., Rondelet (1554), Willughby and Ray (1686), Bloch (1782-1795) and Agassiz (1857), or else as part of a monograph that treated a selected aspect of development, e.g., Tilesius von Tilenau (1802), von Baer (1835), Duvernoy (1844), Van Bambeke (1876), Kingsley and Conn (1883), Henneguy (1888), and Scammon (1911). Comprehensive treatment of fish development was provided in several embryological textbooks, e.g., Balfour (1880), Ziegler (1902), and Kerr (1919).

Access to older, classical literature was provided by Dean's Bibliography of Fishes. It was not until the late Jane Oppenheimer (1936) published her "Historical Introduction to the Study of Teleostean Development" that the topic was subject to historical analysis. Only recently has the history of the study of elasmobranch development been summarized. Little historical documentation has been done for the other groups of fishes.

When considering early studies of fish development, one soon realizes that many of these studies were carried out on viviparous species of elasmobranchs and teleosts rather than oviparous ones, and that both egglaying and livebearing elasmobranchs received a disproportionate amount of attention, considering their relatively low diversity compared to teleosts. First, elasmobranch eggs are considerably larger than those of most teleosts and their development can be easily observed with either the unaided eye or with a hand lens. Secondly, it has always been relatively easy to obtain gravid specimens of viviparous fishes as part of routine fishing efforts.

Moreover, as museum or natural history collections of fishes were made during the course of explorations, specimens of curious, new species of gravid viviparous fishes became available for study, e.g., the four-eyed fish Anableps. In contrast, the eggs of oviparous teleosts are small and not easily collected in the wild. Artificial fertilization of teleost eggs did not begin until the late eighteenth century and the regulated spawning of captive fish species and the rearing of their eggs under "controlled" conditions for scientific purposes was first used only toward the mid-nineteenth century.

In addition, the technology for maintaining fishes in aquaria and the possibility of spawning them under aquarium conditions did not emerge until the mid-late nineteenth century. In passing, one should recall that the Chinese have been rearing captive goldfish, including mutant strains, for over a thousand years. Other factors also influenced decisions to study certain viviparous instead of oviparous fishes. Viviparity in fishes, especially when there is a placental relationship, is an intrinsically interesting process. Finally, as a subject for the general study of development, oviparous teleosts have been secondary to the chicken egg ever since the Renaissance because of the greater size of the chicken's egg and its ready availability as an article of food.

Early History: Aristotle to Bloch

Because the research of early naturalists provided the foundation for the development of fish embryology in the nineteenth century, a brief outline of their work is presented here, beginning with Aristotle. Aristotle, the founder of the science of embryology, was the first to recognize and record many important features of fish reproduction and development (Aristotle, 1942, 1965, 1970). He distinguished oviparous (egg-laying) animals from viviparous (livebearing animals. Aristotle described viviparity in the dogfish shark (Mustelus), the thresher shark (Alopias), and the electric ray (Torpedo). He did not know of viviparity in teleost fishes because there are no viviparous teleosts in the coastal waters of the Mediterranean or the fresh waters of Europe and Asia Minor. His attribution of viviparity to the needlefish, a teleost, is actually a case of misinterpretation. The needlefish of Aristotle is a pipefish or sygnathid. Male pipefishes brood embryos in a pouch of abdominal skin. This form of parental care is analogous to viviparity and in its most highly evolved forms, the parentalembryonic relationships can be as complex as in viviparous fishes.

The body of Aristotle's observations concern chondrichthyan fishes and include: (1) the distinction between oviparous and viviparous modes of reproduction in sharks, rays, and skates; (2) description of the female reproductive system; (3) description of shark and skate egg cases; (4) description of egg structure and some observations on embryonic development; (5) first description of the oviducal or nidamental (=shell) gland but without any knowledge of its function in egg case formation or sperm storage; (6) hatching from the egg case in utero in viviparous species; (7) description of the male reproductive system; (8) notes on breeding seasons and inshore migration of gravid females for "pupping"; and (9) description of the yolk sac placenta in sharks. Writing of the smooth dogfish Mustelus canis in the 5th century B.C., Aristotle states, "the young are produced with the umbilical cord attached to the uterus so that as the substance of the egg gets used up, the embryo's condition appears to be similar to what is found in quadrupeds.

The umbilical cord, which is long, is attached to the lower parts of the uterus; each one is, as it were, fastened to a cotyledon and is attached to the embryo by the middle where the liver is situated. Each embryo has a chorion and membranes of its own round it, just as in quadrupeds. If the embryos are cut open, a situation is disclosed exactly similar to quadrupeds; whatever internal organs they have, such as the liver, are large and supplied with blood".

Aristotle's observations on teleosts are in large part concerned with reproduction, e.g., his exquisite account of brood protection by male catfish. He realized that in contrast to the chondrichthyans, the overwhelming majority of teleosts are oviparous. His observations on the development of teleosts are more limited, due primarily to the small size of most teleostean eggs. Aristotle observed developing teleost eggs, recording their small size and rapid development. He states, "The embryo appears at the upper end of the egg and is enveloped in a membrane; the eyes, large and spherical, are the first organs visible. The embryo and the egg (yolk) are enveloped by a common membrane, and just under this is another membrane that envelops the egg (yolk) by itself; between the two membranes is a liquid". He described the organization of the teleost egg, distinguishing the perivitelline fluid from the yolk mass. [As an aside, Oppenheimer (1936) pointed out that some celebrated nineteenth century embryologists, viz. von Baer and Rathke, erred by considering the perivitelline fluid to be an albumen such as occurs in avian eggs.] Lastly, in comparing the development of teleostean and avian eggs, Aristotle recognized that teleosts lack the allantois.

Relatively little was added to the pioneering efforts of Aristotle by subsequent Greek and Roman scholars. With the decline of classical civilization, the study of science in general as well as the study of fish development languished for almost 1,000 years. Slowly the tide turned. According to Needham (1959), the first account of fish development in medieval Europe was given by Albertus Magnus, a thirteenth century scholastic. During the Renaissance, there was a revival of interest in natural history. The early zoological encyclopedists, especially Belon (1517-1564), Rondelet (1507-1566), and Aldrovandus (1522-1605), investigated fishes and described aspects of their reproduction and development. Again, there was a decided bias toward viviparous chondrichthyans. Rondelet (1554) illustrated an egg case and dissection of the female reproductive system of an oviparous catshark. He also illustrated part of the ovary and an egg case of a skate. He depicted a female shark, probably Mustelus canis, connected to a welldeveloped pup by an elongated yolk stalk that passes from the pup through the cloaca of the female. Although the placenta is not illustrated, Aristotle's research on the shark yolk sac placenta is cited in the text. Progressing to the seventeenth century, Fabricius (1600), who is best known for his studies of the chick, described and illustrated the female reproductive system and uterine embryos with large yolk sacs in a non-placental species of dogfish shark.

By 1673, Niclaus Steno had rediscovered the yolk sac placenta in the smooth dogfish, M. canis, investigated its anatomy, and published an illustrated account of it. He also distinguished between placental viviparity in the smooth dogfish shark, and aplacental viviparity in the spiny dogfish shark, Squalus. In the latter, he discovered that the yolk stalk terminates in an intestine characterized by a spiral valve. His rediscovery appears to have gone unappreciated. During this period, inability to distinguish among species of dogfish sharks that had either aplacental or placental modes of viviparity caused the very existence of a yolk sac placenta in sharks to be doubted. This controversy was not fully resolved until the time of Johannes Muller. Steno's studies on shark development had important ramifications for the study of mammalian reproduction. On the basis of his observation of large eggs in the uterine oviduct and large eggs in the ovary of the spiny dogfish, he concluded that the embryo does not form de novo in the oviduct, but comes form an egg released by one of the ovaries. From this, he postulated that mammalian development proceeded in a similar fashion, except that the mammalian egg is much smaller. In his examination of the mammalian ovary, unfortunately, he mistook the entire follicle for the mammalian egg. These studies were undertaken at a time when a controversy had arisen between animalculists and ovists with respect to the primacy of the sperm in reproduction and the absence of any knowledge of the mammalian egg. Steno's observations, although flawed, brought mammlian developmen in line with that of other vertebrates. In essence, he used sharks as a "model" system for mammals.

In 1678, Stephen Lorenzini, a pupil of Steno and Redi, published a set of observations on the viviparous electric ray Torpedo. This appears to be the first monographic treatment of a single fish species. Lorenzini's work is of considerable interest because almost one third of it is devoted to an account of urogenital anatomy, reproductive biology, and development of Torpedo, as well as several other rays.

Amongst his observations, those of primary interest concern oogenesis and ovulation, the blastoderm, and embryonic nutrition. Describing the ovary, he notes differences in size and structure between young and mature fishes. Maturing oocytes and ova vary in size, composition, and colour. Oocytes are connected to the ovarian matrix by a stalk through which several blood vessels pass, an arrangement similar to that described by Fabricius in the hen's egg. He states that these vessels provide nutrients to the eggs. The maturing egg is invested by a "membrane" except for the pole opposite the site of attachment to the stalk.

This pole constitutes the pore or stigma through which the mature egg is released from the follicle. During oocyte maturation, changes in membrane adhesion were documented. Lorenzini speculates that mature eggs are released from the follicle by contraction of fibres in the ovarian stroma. The ovulated egg is "received" by the oviduct and passes into the uterine region where embryonic development is completed. Following Fabricius (1537-1619), De Graaf (1641-1673), and Harvey (15781657), Lorenzini refers to the residual portion of the follicle as the "calyx." Calyces, which apparently are the corpora lutea, occur in various sizes and shapes throughout gestation, and are lost or regress some time after parturition. Lorenzini found a correlation between the number of young in a brood and the number of calyces in the ovary. Ovulation in the Torpedo is considered to conform to De Graaf's account in other species. Among the more interesting of Lorenzini's observations are those made on the early stages of embryonic development. He describes and illustrates two neurula stages. The earlier stage depicts an early neural groove on the surface of the blastoderm. He describes the characteristic chondrichthyan blastoderm, "a leaden gray disc" that is clearly distinct from and surmounts the "yellow yolk mass." In another egg, he describes and depicts a slightly more advanced neurula stage (equivalent to Ballard et al., 1993, Stage 15). Here Lorenzini correctly shows the splayed, outspread cephalic margins of the medullary plate, a characteristic feature of chondrichthyan embryos. Lorenzini also observed the blastoderm in three other species of rays.

Lorenzini seems to be the first to have depicted the chondrichthyan blastoderm. His observations appeared only six years after Malphighi's initial account of the avian (chick) blastoderm (Needham, 1959). Lorenzini made one of the first major contributions to the knowledge of embryonic nutrition in non-placental viviparous fishes. Late stage embryos with yolk sacs lie free in the uterine region of the oviduct where they are bathed by uterine fluid. A yolk stalk passes from the external yolk through the abdominal wall and empties into a large internal yolk sac that in turn empties into the intestine.

The intestine is filled with yellow material apparently derived from the yolk and brown globules. The same brown globules occur in the uterine fluid and in the mouth, esophagus, and stomach of the embryos. Lorenzini concluded that advanced embryos are nourished both by yolk passed to the intestine and by uterine fluid which they imbibe. He believed that the nutrients in uterine fluid are derived from the blood and conveyed to the uterine fluid by vascularized glandular structures which he compared to chorionic villi. These structures are now known as trophonemata. Villi increase in size during gestation; their size is correlated with the amount of uterine fluid. They regress and are absent in non-gravid fish. In addition, Lorenzini also made significant observations on the male reproductive system, reproductive cycles, and the duration of gestation. In the closing years of the seventeenth century, Willughby and Ray (1686) summarized much of what was known of fish reproduction and development in a work primarily devoted to systematics.

Considering its promising beginning, it is surprising that the study of chondrichthyan development made relatively little progress in the eighteenth century. There are many possible reasons. Those scholars interested in fis biology may have been attracted to systematic studies because of the revolutionary efforts of Willughby and Ray, Artedi (1705-1735), and Linnaeus (17071778) as well as the influx of new species of fishes from the tropics. Changing fashions in science also enter the picture, especially the rise of microscopy and microscopic anatomy. Perhaps in terms of the material, questions, and technology, research on chondrichthyans had reached an impasse.

There are some exceptions. Bohadsch (1776) investigated the structural organization of eggcases. Monro (1785) published detailed descriptions accompanied by elegant illustrations of the urogenital anatomy of the oviparous skates. He also described and illustrated for the first time the spermocysts, spherical aggregations of Sertoli cells and differentiating sperm cells that are a characteristic feature of chondrichthyan spermatogenesis. Descriptions and illustrations of microdissections of the internal anatomy of mid-and late-stage embryos are given. The extensive vitelline circulation and the connection of the yolk sac and the small intestine are documented. External gill filaments, another feature of chondrichthyan development, are described for the first time. The eighteenth century concluded with Bloch's magnum opus (1782-95) which added little to the study of chondrichthyan development other than information on the embryos of the viviparous sawfish Pristis.

The study of teleostean development, as previously noted, lagged well behind that of both oviparous and viviparous chondrichthyan fishes, and little was done prior to the mid-late eighteenth century. To be more accurate, the little that was done was done with viviparous teleosts. Schoenveld (1624) first described viviparity in the embryos of the eelpout, Zoarces viviparus, a fish common to the marine waters of northern Europe. Late stage embryos are illustrated. Impressed by the substantial growth of the embryos during gestation, Schoenveld postulated that they obtain nutrients by ingesting ovarian fluid. (In viviparous teleosts, an oviduct or uterus is absent and gestation takes place in the ovary. Cf. Wourms, et al., 1988). Somewhat later, Willughby and Ray (1686) apparently observed the development of live Zoarces embryos. Interest in teleost development during the seventeenth and eighteenth centuries seems to have been generated by the viviparity of exotic species discovered during the explorations of the Old and New World Tropics.

Anableps, the four-eyed fish, conspicuous because of its size, unusual anatomy, and behaviour in its Central and South American habitats, was described as viviparous as early as 1738. Gronovius (1754-1756) was the first to illustrate advanced embryos of Anableps with their characteristic rugose "yolk sac" (actually the abdominal trophoderm). Gronovius may also have been the first to describe viviparity in the clinid, Clinus superciliosus. Bloch (1782-1795) described and illustrated the external and internal reproductive anatomy of Anableps, its embryos in situ, and the embryos themselves. He also depicted the adults and embryos of Zoarces and Clinus. One of the first sets of field observations was made by de Alzate — Ramyrez in 1769 who discovered viviparity in a Mexican poeciliid and made microscopic observations of blood circulation in living, near-term embryos.

During this period, the foundations were laid for the study of early development, i.e., fertilization through embryogenesis. In 1763, Jacobi (cited in Oppenheimer, 1936) published an extensive letter in which he reported that he had artificially fertilized the eggs of trout and salmon by removing the ripe eggs and adding sperm to them. Although this is the first known written account of artificial fertilization, it is probable that the technique had been in use for some time, for fish culture had been practiced in Europe since the late fifteenth century. Jacobi may have observed the entry of sperm into the egg. He also recorded the formation of double embryos in artificially fertilized eggs, a condition that he attributed to the entry of two spermatozoa. In 1787, Cavolini published a monograph on the "generation" of fishes and crabs. The bulk of this work was devoted to the description of spawning habits, hermaphroditism, fertilization, and contemplations on the nature of development. However, he did make one observation of considerable significance. In the eggs of the pipefish, he described a structure on the egg which he referred to as the "cicatrix" or in modern terminology, the blastoderm.

He identified the blastoderm with the same structure in the chick and assumed that the embryo originated from the blastoderm. The study of teleost development in the eighteenth century was brought to a close with the appearance of Bloch's (1782-1795) massive, twelve-volume monograph on the natural history of fishes. This work is one of the linchpins of systematic ichthyology and summarized what was known about fishes up to that time. It contains much information on fish reproduction and development. As previously noted, aspects of the development of viviparous chondrichthyans and teleosts are described and illustrated.

Of greater interest are his pioneering studies of the early stages of development of an oviparous teleost. For the first time, he describes and illustrates consecutive stages in development, from post fertilization through hatching to a late larval stage, and provides a time base for the description of developmental stages. He appreciated the eccentric position of the yolk mass relative to the blastoderm. He made observations on the development of the heart, and illustrates the contractile, simple tubular stage of heart development. There were, of course, errors of interpretation, one of which was the identification of the perivitelline fluid of the teleost egg with the albumen layer of the avian egg.

Development of Teleostean Fishes

As the nineteenth century dawned, the foundations had been laid for the study of fish development, and the stage had been set for the extraordinary progress that would soon be made. The rise of fish embryology was vitally linked to the origins of embryology as a formal science and its subsequent growth, diversification, and interaction with other branches of biology. Scientific embryology is generally considered to begin with the publication of the first volume of von Baer's comparative studies of animal development. As Oppenheimer (1936) has pointed out, even though von Baer's work codified embryology as a science and profoundly influenced its future directions, nonetheless, it was the culmination of a course that had been pioneered by Dollinger (1770-1841) and Rathke (1793-1860). Oppenheimer (1936) states that Dollinger "was the originator of the appreciation of comparative development by working out carefully the succession of steps in the developmental processes in various vertebrate forms." Dollinger not only influenced von Baer (1792-1876) but also guided Pander's (1793-1860) revolutionary studies of chick development and inspired in Louis Agassiz (1807-1873) a life-long commitment to a program of comparative embryological research. Dollinger's approach to embryology had its origins in the influence of German Naturphilosophie on the study of comparative anatomy. Thus influenced, anatomists attempted to systematize, describe, and classify all organisms or parts of organisms as variations of a single type. Dollinger, who was a proponent of Naturphilosophie, thus accepted and extended suggestions of the anatomist Meckel (1781-1833) that a unity of plan manifests itself in development as well as in structure.

For reasons of convenience and also because their study progressed at different rates and in different ways, I have chosen to treat teleosts and chondrichthyans separately. In retrospect, it is amazing to contemplate the almost explosive increase in knowledge of teleostean development that took place in the period 1819-1842. The first major work from this period was Forchhammer's (1819) dissertation on the development of the viviparous blenny Zoarces. He described no stages younger than pre-motile embryos.

The main body of the work comprises descriptions of older embryos and organ formation. He was the first to describe the embryonic shield, the region in which the embryo forms. His observations on the contractility of the excised embryonic heart constitute one of the earliest attempts to study isolated embryonic organs. He also observed that yolk ultimately is absorbed into the liver. Rathke (1833) also published a monograph on the development of the viviparous blenny, as part of a much larger serial work on the development of animals and humans. Conceptually, Rathke's studies parallel those of Dollinger and von Baer (Oppenheimer, 1936). Although he described and analysed early stages of development, he was primarily concerned with older stages. His observations on the developmental anatomy of the brain, heart, gut, and other organs established a new standard of excellence in the field.

The first volume of von Baer's three-volume treatise on embryology appeared in 1828 but fishes were not dealt with until the second volume. In the interim, he published a separate communication devoted to the development of fishes based on his studies of species of Cyprinus. He made several fundamental observations such as the sharp distinction between the pattern of development in large-yolked and small-yolked teleostean eggs and the great differences in the duration ofdevelopment, viz. three months in trout, a few weeks in Zoarces and a few days in Cyprinus. His choice of Cyprinus, an egg-laying species, made it possible to obtain newly fertilized eggs, thus giving him access to the early, i.e., cleavage-epiboly, stages of development and providing a chronological base for the consecutive stages of development. von Baer (1835) was the first to describe clearly and illustrate one of the most characteristic features of teleostean development, viz. epiboly or overgrowth of the yolk by the blastoderm. Not only did he describe it, but also he provided a timetable for the process. He was also the first to describe the germring, i.e., the thickened edge of the blastoderm. von Baer failed to observe cleavage and incorrectly homologized teleost development with that of other vertebrates by mis-identifying the early embryonic shield as primitive streak and the late shield as a broad neural plate. von Baer concluded his monograph on fish development with an appendix in which he discussed the embryological origin and function of the swimbladder and compared it to the lungs of other vertebrates.

Studies very much similar to those of von Baer were being carried out by Rusconi (1836) on two species of Cyprinus. Rosconi gives a detailed account of his method of artificial fertilization. Developing eggs, thus obtained, were of known age and were subsequently examined microscopically. His major contribution is that he appears to have been the first to describe cleavage in the teleostean egg. His detailed description is illustrated by a plate which "rivals all other in the literature for accuracy and beauty in presenting the early stages of teleostean development".

He also made extremely accurate observations on the embryonic shield and its role in the formation of the embryo. To aid observation, he used an acid to stop development and render the embryo opaque without affecting the transparency of the yolk. His figures depict the concentration of cells in the midline of the embryonic shield that are the precursor of notochord and spinal cord. Elsewhere, he states for the first time that the teleostean nerve cord originates from a single structure rather than by the folding of a neural plate.

Embryological research on teleosts moved to France during the middle years of the century. The first important work was Vogt's (1842) contribution to the embryology of the whitefish (Coregonus), which was part of Agassiz's elaborate series of monographs on the freshwater fishes of central Europe. The embryology portion consists of a text volume and a folio of coloured lithograph plates. Although considered part of the French School by Oppenheimer (1936), it could just as well be considered a product of the Dollinger-Wurzburg School.

Agassiz was in close contact with Dollinger at Munich between 1827 and 1830 and was studying teleost development. The role of Agassiz, if any, in Vogt's study is not clear, for the dispute over authorship was the first in a series of controversies that would occur between Agassiz and his collaborators. In addition to its scientific methods (vide infra), this work had an extraordinary impact because of the clarity and beauty of its illustrations. The figures were drawn by professional artists and reproduced, many in colour, by lithography, a printing process capable of rendering fine detail. Not only were new observations effectively presented, but also the known details of development took on a new lucidity. The Vogt-Agassiz study is another study in which developing eggs were obtained by artificial fertilization, cultured within a set temperature range (5-IOC), and examined microscopically For the first time, the technique of dechorionation is mentioned as an aid for observation. Vogt also used the technique of acid fixation to enhance the visibility of the embryo. A series of well-illustrated, closely spaced stages in the normal development of the whitefish was presented. This work appears to be the first study to apply a cellular approach to the interpretation of developmental events, especially oogenesis, cleavage, epiboly, and embryogenesis. Information is provided on oogenesis, the structure of the mature egg, and fertilization.

According to Oppenheimer (1936), Vogt was the first to inquire into the mechanics of embryo formation. He also speculated on the role of cell movement and cell migration in development. A considerable portion of the work was devoted to the microanatomy of the later stages of embryonic and larval development. In this respect, Vogt gives an account of the formation of a number of organ systems, tracing their development from the earliest cellular condensations or anlagen to completion. Illustrations of melanocytes and the cellular components of the notochord and connective tissue are remarkable for their detail which included nuclei.

The next figure of consequence in the French school was Lereboullet, a pupil of Duvernoy, who in turn was a protege of Cuvier. Lereboullet made substantial contributions in the field of descriptive embryology and subsequently in experimental studies which were the first great attempt to analyse the mechanism of embryo formation. His monograph on the development of the pike and perch established new standards in terms of the accuracy with which his observations were depicted. He provided the first detailed description of the microscopic organization of the egg and a more accurate description and illustration of the embryonic shield. He stated that it is the precursor of the entire embryo and not just the central nervous system. His most important single observation was the description of the yolk syncytial layer (periblast) which he termed "feuillet muqueux," literally mucous lamina.

He mistakenly ascribed to it the fate of forming elements of the digestive system. He regarded the germ ring, i.e., the thickened margin of the blastoderm, as a dynamic force during the formation of the embryo. This viewpoint, reinforced by his studies n experimental teratology (vide infra), led him to interpret the process of embryo formation in terms of concrescence, i.e., the right and left portions of he germ ring assemble into the two halves of the body. Lereboullet arrved at this conclusion a decade before His (1874) formally promulgated the concrescence theory. Finally, in his monograph on the comparative embryology of the trout, a lizard and a snail, he refuted the concept of a unity in the plan of development propounded by Meckel and Serres. Three important structures, characteristic of teleostean development, still need to be dealt wit, namely the yolk syncytial layer or periblast, the enveloping layer, and Kupffer's esicle. Although Lereboullet (1862) first described the yolk syncytial layer, it was vo Kupffer (1868) who suggested that the cells of the blastoderm not only give rise to the tissue cells of the embryo but also are progenitors of he yolk syncytial layer. Agassiz and Whitman (1884, 1915) proved beyond any doubt that yolk syncytial nuclei are derived from the marginal cells o the blastoderm.

The enveloping cell layer is a tightly sealed epithelial monolayer that forms the outermost boundary of the blastoderm. Priority for its discovery is unclear, for descriptions of it under various names appeared more or less simultaneously in a number of indepndent studies. Kupffer's vesicle is a transient structure that is only found in teleost embryos. It was discovered by von Kupffer (1866), who thought that it might be an allantoic rudiment. It is a small, fluid-filed epithelial sac that appears during the somite stages. Kupffer's vesicle develops mid-ventrally in a region below the tail bud and at the interface between the embryo and yolk mass. Even now its function, if any, still remains unknown.

At this point, a digression to consider the development of viviparous teleosts is in order. Following the pioneering studies of Forchhammer (1819) and Rathke (1833) on Zoarces, knowledge of viviparity in teleosts progressed steadily through the rest of the century. In 1844, Duvernoy escribed the development of a freshwater poeciliid fish from Surinam. His work was mainly concerned with the later stages of development, but did provide some important information on early development. The ovary is the site of gestation. Fertilization takes place in the follicle where the embryo is invested by an egg envelope. Development to term takes place in the follicle. Duvernoy illustrated for the first time an early stage follicular embryo in its egg envelope. In his Memorias, published 1851-55, the Cuban naturalist Poey was able to extend the number of viviparous species to include all ten species ofCuban poeciliids.

Of greater significance were his observations for the first time of sperm storage in these fishes. Unfortunately, he interpreted his observations in terms of parthenogenesis rather than sperm storage and superfetation. The problem was not satisfactorily resolved until 1908, when Phillipi discovered that in poeciliids, sperm is packaged into spermatozeugmata that are transferred to the male reproductive tract. He also presented evidence for specific structural pathways, which he called "delle," that allow sperm to move from the lumen into close proximity with the egg and fertilize it within the follicle. In 1856, Wyman produced a definitive study on the embryology of the foureyed fish Anableps. It, too, is a fish with follicular gestation. He depicts the elaborate follicular epithelium and the highly vascularized abdominal trophoderm of the embryo. Because of the massive increase in size of the embryo during gestation, he postulated that the developing embryos are nourished by secreted maternal proteins. He extended his speculations to include embryonic nutrition in the surfperch and the electric ray Torpedo.

In 1853, Agassiz first described viviparity in the surfperches or embiotocids from the Pacific coast of North America. In a more extensive publication, Girard (1858) described and illustrated 17 species. He provided the first detailed information about their reproduction and development. Gestation is ovarian: embryos develop in the ovarian lumen and are enfolded by sheaths of ovarian tissue. Initial egg size is minute, but term embryos are large (5 cm). A series of developmental stages is described and illustrated. Figures on brood size which varies (10-80 young) according to species are given. The duration of gestation is estimated at 5-6 months.

The initial descriptions of Agassiz and Girard culminated in a definitive study of surfperch development by Eigenmann (1894) which is tour de force of descriptive embryology. He noted that the early cleavage stages of the small, yolkpoor eggs verge on being holoblastic. He described the presence of sperm in the ovarian fluid and the imbibition of ovarian fluid by the embryos. Male embryos undergo precocious sexual development. In an earlier separate publication, he described the precocious segregation of germ cell determinants at an early cleavage stage, an overlooked, early contribution to concepts of germ cell development.

Throughout the remaining years of the century the occurrence of viviparity was described in various groups of teleost, e.g., the rockfish, Sebastes, halfbeaks, Hemirhamphus, and the Lake Baikal oilfish, Comephorus. One should recall that such studies are still continuing, with viviparity in the living coelacanth Latimeria not discovered until 1975. By 1886, the field had progressed to the extent that Ryder (1886a) could review it and add to it some of his own original observations.

He concluded that the developing embryos of surf perches derive nutrients by absorbing imbibed ovarian fluid through elongated intestinal villi found in the hypertrophied hindgut. He also concluded that the other characteristic feature of surfperch embryos, the unique vascular apparatus of the enlarged vertical fins, serves a respiratory rather than a nutritive function. Much of the research on viviparity, both in teleosts and chondrichthyans during the latter part of the century was concerned with the elaboration of highly diverse maternal and embryonic structural adaptations for viviparity and attempts to correlate these structures with physiological function, e.g., nutrition, gas exchange. Advances in technology had profound effects on embryological research during the latter half of the century. Because of the transparency of most teleostean eggs observations had usually been made on intact eggs or microdissections of embryos. Beginning in 1860, a series of discoveries and innovations in histological microtechnique made it possible to fix, embed, section, stain, and subsequently examine serial sections of embryonic material with microscopes equipped with high resolution, apochromatic, oil-immersion objectives (cf. Ryder, 1884b and Whitman, 1885 for the availability of tehniques in embryology, and Horder et al., 1985, for a chronology of technological and conceptual innovations). In this respect, the application of photography to the study of fish development occurred surprisingly early in the history of photography. Von Kupffer is credited with the first extensive use of photomicrographs to illustrate an embryological monograph in his 1878 study of herring development. By 1898, His had published micrographs of sectioned trout blastomeres that illustrated aspects of chromosome behaviour during cell division.

The consequences of these advances in embryological technology was to trigger a veritable explosion of embryological studies of teleosts based on the microscopic study of sectioned material, e.g., Van Bambeke, 1876; Hoffman, 1881; Henneguy, 1888; Ryder, 1884a; Kowalewsky, 1886; and Wilson, 1891. By the end of the nineteenth century, an excellent descriptive knowledge of teleost development had been achieved. But there was a trade-off. The emphasis on the study of sectioned material detracted from the observation of living embryos, especially cell behaviour during early development, a field that was just beginning to be explored by His (vide infra). This trend led to subsequent difficulties in the interpretation of morphogenetic movements and embryo formation. Even as this trend was established, its advantages and limitations were becoming obvious. Harrison (1901) published his histological study of the differentiation and axonal growth of spinal neurons in trout embryos in which he focused on the patterns of growth of individual nerve cells and their role in establishing neural connections. The constraints of the histological aproach led him to develop tissue culture and use it six years later to investigate neuronal outgrowth.

Development of Chondrichthyan Fishes

At the onset of the nineteenth century, the study of chondrichthyan development was already farther advanced than that of teleosts. However, the very features that had initially favoured the study of chondrichthyan fishes, e.g., large egs and embryos, viviparity, and internal fertilization, subsequently worked against their further study. It was difficult to obtain access to early stages of development in viviparous species and there was no really effective way to maintain egg-laying species in captivity so as to obtain newly laid eggs. The rate of development was slow. The large, meroblastic eggs were opaque, so that only external features of development and those details revealed by microdissections of embryos could be observed.

In contrast, the introduction of artificial fertilization ade the early stages of teleostean development readily available and the transparency of most teleostean eggs and embryos was well suited for the microscopic examination of external and internal features of development. Accordingly, nineteenth century studies of chondrichthyan development can be divided into two phases. Phase One spanned the period 1800-1860. During this phase, research progressed along previously established lines with an emphasis on reproductive biology, the study of viviparous species and adaptations for viviparity, and the later stages of embryonic development. Phase Two spanned the period 1860-1911. Research in this phase was shaped by the introduction of new embryological and histological microtechniques to the study of chondrichthyans.

Research of the eighteenth century passed imperceptibly into the nineteenth century with little or no change of agenda. In addition to his magnum opus, Bloch published a set of separate papers in obscure journals, some of which dealt with chondrichthyan reproduction and development. Tilesius von Tilenau (1802) published a monograph on the egg cases of sharks and skates. It is noteworthy because it represents the first attempts to determine the chemical composition of egg cases, now known to be a type of "tanned" protein. In addition, he also reported on the urogenital anatomy and the reproduction and development of sharks and skates. The major significance of Mitchell's (1803) report on the anatomy of early (7-8 cm) pre-implantation shark embryos is that it is one of the earliest embryological papers and probably the first paper on fish embryology published in the United States.

Home's (1810) paper had two major aspects: one embryological and the other reproductive. He provided anatomical descriptions of the male and female reproductive system and described the structural changes that occur as mature individuals pass from a reproductively inactive to an active state. His study of the male urogenital system and claspers (modified caudal portion of each pelvic fin) represents the first step toward understanding the process of chondrichthyan insemination and fertilization. He gives a fine illustration of the claspers with their terminal processes expanded as they would be in copulation. He also describes the urogenital papilla, the terminal end of the male reproductive duct, but mistakenly refers to it as a "penis" and states that it is the intromittent organ.

Claspers were first reported by Aristotle, and their gross anatomy described by Lorenzini (1678) and Bloch (1782-1795). A controversy developed over the centuries concerning their function either as an intromittent organ or an organ for holding the female fast during copulation Further contributions were made by Davy (1839). He described the internal anatomy of the clasper that makes possible its function in sperm transfer. On the basis of his own work and a review of the literature, he concluded that the clasper functions both as a holdfast and intromittent organ. He also described the accessory clasper gland of skates and rays, a structure now known to produce prostaglandins (Lacy, personal communication).

Davy (1839) appears to be the first to have observed chondrichthyan sperm. In the electric ray (Torpedo) and a skate, he described sperm with an elongated head and flagellum. In vitro microscopic examination revealed that their motion was "serpentine and vibratory." Details on the process of spermatogenesis and the structure of the mature sperm came later. In 1871, Agassiz stated that when the claspers (of a skate) are rotated forward and upwards, an opening in them was brought opposite the spermatic duct (=urogenital papilla).

The clasper could then be introduced into the oviduct and spermatic fluid could pass up the clasper and into the oviduct. Bolau (1881) appears to be the first to actually observe and describe the mating activity of sharks. The current view of insemination is that of Gilbert and Heath: "In mating, one clasper is flexed medially, inserted, and is anchored in the oviduct by a complex of cartilages at the clasper tip. Sperm pass from the urogenital papilla into the clasper groove and are washed into the oviduct by seawater and secretions expressed from a siphon sac." The actual cytological events of fertilization in chondrichthyans did not become known until Ruckert (1892b) demonstrated physiological polyspermy.

Embryology was the second aspect of Home's (1810) paper. He compared stages in development of the oviparous catshark, Scyliorhinus with the non-placental viviparous spiny dogfish shark, Squalus. He illustrated an advanced yolk sac embryo of the catshark in its egg case. He illustrated and described the slits that open in the egg case to permit the flow of seawater through the egg case during mid-late phases of development. He gave an excellent rendering of an oviduct of gravid Squalus with its ostium tubae, nidamental (shell) gland, and posterior uterus lined with longitudinal vascular ridges. Within the uterus, the "candle" is illustrated for the first time.

It is an attenuated egg case that contains three developing eggs and a quantity of transparent jelly. Neonates of both the catshark and spiny dogfish are illustrated. To determine the physical and cemical properties of the oviducal jelly of the spiny dogfish, Home (1810) enlisted the aid of the chemist W. Brande. The jelly was extremely hydrophilic, being able to absorb many times its volume of water. Chemically the jelly was found to be neither gelatin nor albumen. It is interesting to note that Home, under the influence of Humphrey Davy, speculated on the requirements of developing fish eggs for aeration and how this process might be accomplished with different patterns of embryonic circulation.

Rathke was von Baer's successor at Konigsberg. His study of the "sexual apparatus" in fishes and other vertebrates was the first major step in understanding the development of the urogenital system, the differentiation of the gonads and accessory ducts, and the clarification of relationships in the vertebrate urogenital system. His monograph on the development of sharks and rays was primarily concerned with organogenesis and the microanatomy of the developing embryo. His comparative study of the development of the gill slits and arches forms the basis for homologizing the pharyngula phase of vertebrate development and was used extensively by the proponents of the biogenetic law later in the century.

John Davy, the brother of Humphrey Davy, published an account of the embryonic development of the electric ray Torpedo and a series of experiments on its electric organs. In his description of the female reproductive system, he noted that the absence of a nidamental (shell) gland accounts for the absence of an egg envelope in this fish. He presents an illustrated series of developmental stages which includes information on the electric organ. His earliest embryo was 18 mm long. It had short, external gill filaments but lacked an electric organ.

The electric organ had begun its development in 28 mm embryos. Pectoral fin development began in 39-35 mm embryos, and by 62 mm the embryo looked like a miniature adult. Living full-term embryos were surgically removed from the female and maintained in sea water for up to six months. These neonates had functional electric organs. Activity was measured with a galvanometer. At six months, the young rays still had vestiges of an internal yolk sac and showed no interest in food. Davy suggests that they depend on yolk reserves during this period. Davy's (1834) contributions to the study of embryonic nutrition will be discussed in another section. Almost simultaneously, Leuckart (1836), a protege of Rathke, published a comparative study of the external gill filaments of several species of sharks and the electric ray.

This paper has added interest because Leuckart's incorporation of an extensive series of measurements of embryonic structures represents an early attempt to quantify embryological research. Based on his own work and that of others, Leuckart concluded that external gill filamets are transitory structures that are common to the embryos of all chondrichthyan fishes. He suggested that passage of the embryos through an external gill filament phase is a type of metamorphosis.

In addition, for the first time he describes the yolk stalk appendiculae of the pre-implantation embryos of the bonnethead shark, Sphyrna tiburo. They are vascularized villiform projections that festoon the umbilical cord of this placental species. Johannes Muller was one of the pivotal figures in mid-nineteenth century biology as a pioneer marine biologist, a founder of comparative physiology, an accomplished embryologist, and a mentor of illustrious students such as Henle, K6lliker, Haeckel, His, Virchow, and Du Bois-Reymond. He helped to lay the foundations for the experimental approach to embryology. Muller's classical paper of 1842 is considered a benchmark of chondrichthyan research. In it, he employed a comparative approach to the study of oviparous and viviparous species. He described and compared the egg cases of sharks, skates, and chimaeras. More importantly, he provided a detailed description of the yolk sac placenta in the smooth dogfish shark, Mustelus canis and the blue shark, Prionace glauca.

He gave an in-depth account of the maternal and fetal portions of the placenta and their vascularization. Prior to Mller's definitive account, knowledge of the shark yolk sac placenta had languished since Steno's (1673) rediscovery of it, over 170 years previously. Shortly thereafter, Leydig's (1852) monograph on the histology and embryology of rays and sharks appeared. In it he described five stages in the early development of the spiny dogfish, Squalus acanthias. A detailed description and accurate illustration of the microanatomy of the earliest stage (equivalent to Stage 23 of Ballard et al., 1993) is given. His research on the yolk sac placenta extended that of Miller. The embryonic portion of the placenta of M. laevis at its attachment site is illustrated and a schematic illustration of the histology of the maternal embryonic placental interface is provided. In another advance, Leydig (1852) described the differentiation of six tissues, viz. neurons, notochord, connective tissue, lens fibres, heart muscle, and striated muscle. In the latter tissue, cell fusion was depicted.

Although the eggs and embryos of viviparous species continued to be used for the study of development, the study of viviparity per se entered a period of stasis that lasted well into the twentieth century. However, some progress was made. Bruch (1860) published the first accurate descriptions and illustrations of the uterine villi of several viviparous rays. The extensive vascularization and glandular oganization of these structures were noted. These observations were confirmed and extended by Trois (1876). Wood-Mason and Alcock (1891) discovered that the uterine villi of the butterfly ray (Gymnura) are grossly hypertrophied in one region of the uterus and enter the spiracles of the embryo.

This juxtaposition of maternal and embryonic tissues was subsequently termed "a branchial placenta". WoodMason and Alcock coined the term "trophonemata," literally growth threads, to describe the hypertrophied uterine villi. They ascribe a function of nutrient production to the trophonemata. Developing embryos would imbibe the nutrients. Although studies of the yolk sac placenta also lagged, Ercolani's (1879) observations on shark placentae were incorporated into his general classification of the types of vertebrate placentae. Parker (1889) described the fetal membranes of Mustelus antarcticus and provided an analysis of the periembryonic fluid. Alcock (1890) described the histology of the spatulate extensions that occur on the umbilical cord of the yolk sac placenta in the hammerhead shark Sphyrna blochi. He uses, apparently for the first time, the term "appendiculae" to categorize these processes. Their function was enigmatic. Alcock speculates on a possible role as a lymphatic gland or involvement in the "purification" of embryonic blood.

Advances in embryological microtechnique and microscopy that so profoundly altered the study of teleostean embryology had an even more pronounced effect on the study of chondrichthyans because of their large, opaque, meroblastic eggs. Research shifted to the histological and cellular levels. The consequences are especially apparent in research on early development, i.e., fertilization through early development, as well as organogenesis and cell differentiation.

Although the blastoderm and neurula stages had been recognized by Lorenzini (1678), little progress had been made in the intervening years in defining the blastula and gastrula stages. Coste (1850) appears to be the first to describe but not illustrate cleavage. His report was extended by Gerbe (1872) who illustrated surface views of stages from the first cleavage through a late blastula and compared them to the cleavage stages of birds and reptiles. These observations were confirmed by Balfour (1878) who also included illustrations of sectioned blastulae. The definitive study of early development from fertilization through the early gastrula was conducted by Ruckert (1899). Surface views of the blastoderm were accompanied by sections through it.

In 1878, Balfour published his monograph, On the Development of Elasmobranch Fishes. This work raised the study of chondrichthyan development to heretofore unattained heights by establishing new standards of excellence for accuracy and comprehensive description. It has been categorized as "a famous landmark in the history of vertebrate development". Balfour explored all aspects of elasmobranch embryology from early cleavage through mid-late development and established a staged series of embryos, mostly sharks, that was employed in research for many years. Illustrations of sectioned material were used extensively. Balfour described the organization of the mature egg and cleavage of the zygote, commenting on yolk platelets, the germinal vesicle and its fate, the process of cleavage, nuclear division, the origin of yolk nuclei, and asymmetry of the blastoderm.

An indepth account of gastrulation, germ layer formation, and the general details of embryonic development is presented. Detailed descriptions of the development of the major organ systems included: (1) brain, spinal and sympathetic nerves; (2) head and associated structures; (3) digestive tube and associated organs; (4) heart and vascular system; and (5) gonads, kidney,

Wolffian and Mllerian ducts and their derivatives. In passing, it should be noted that Balfour conducted much of his research at the Naples Zoological Station. His research was facilitated by the director, Anton Dohrn, who was pursuing a major research program in evolutionary physiological anatomy. This program included studies of shark embryology by Dohrn and his associates in order to resolve Dohrn's ideas about vertebrate origins.

Observations by Ziegler and Ziegler (1892) on the development of the electric ray Torpedo continued Balfour's pioneering approach. Their work is characterized by the use of serial sections and photographs of surface views of model reconstructions of selected embryonic stages. They investigated the equivalent of Balfour stages B to K, and considerably advanced descriptive knowledge of gastrulation, formation of the medullary plate, closure of the medullary plate and early embryo formation, and the organization of gill slit-tail bud stage embryos. Balfour had made major advances in the study of chondrichthyan development and the introduction of a series of normal stages was useful. However, there were defects in his norma series. The series was intended to illustrate development in Scyliorhinus, but it was necessary to use embryos of the shark Galeus and the ray Torpedo to fill in gaps.

Moreover, there was no time base to the series. As part of Keibel's series of monographs on the normal stages of development of vertebrates, Scammon (1911) published his monograph on the spiny dogfish shark, Squalus acanthias. It is a reasonably complete survey of the development of a single elasmobranch species. Ballard et aL (1993, p. 334) refer to Scammon's monographs as "hitherto unrivaled in its completeness, records in handsome drawings, wax reconstructions and serial sections the anatomy of dozens of randomly collected but closely spaced specimens." Scammon (1911) also included a bibliography arranged by subject of the important literature on chondrichthyan development. The appearance of Scammon's monograph represents the apex of nineteenth century studies of chondrichthyans.

A major conceptual difficulty encountered in the study of chondrichthyan development is the lack of correspondence of the processes of gastrulation and embryo formation with those of other vertebrates with meroblastic eggs, in particular, teleosts and reptiles and birds. Suffice to say that the posterior rim of the chondrichthyan blastoderm thickens to form a medial embryonic shield with two lateral crescent-like arms. The arms eventually close to form the embryonic axis. The embryo-forming region overhangs the yolk mass. Initially, the anterior end of the embryonic axis is a fixed point and lengthening of the axis is accomplished by posterior growth brought about by the joining together of the right and left arms of the shield Although considerable effort has been invested in the descriptive and experimental study of the problem, according to Ballard et al. (1993, p. 328), "No consensus has been reached as to how the morphogenetic cell movements are taking place in this or any other elasmobranch fish." Another area of inquiry in which the advances in embryological microtechnique provided spectacular insights was the study of oogenesis. Although lampbrush chromosomes were first seen by Flemming in 1878, it was Ruckert (1892a) who first gave evidence that these structures are chromosomes. He investigated the fate of the chromosomal material during the growth of the oocyte nucleus in the ovaries of three elasmobranchs, Scyllium (=Scyliorhinus), Torpedo, and Pristiurus (=Galeus). Ruckert made three major contributions.

He was the first to devise a method of isolating intact germinal vesicles and staning lampbrush chromosomes in situ. Secondly, he described the characteristic structure of the lampbrush chromosomes, especially the lateral loops, and was the first to use the term "lampbrush." Finally, by examining stages in oogenesis, he described the genesis of lampbrush chromosomes during early stages of oocyte growth (18 wm to 2 mm), their unique structural configuration at full development (2-3 mm), and the retraction of the lateral loops and contraction of the chromosome axis as these chromosomes give rise to normal condensed meiotic bivalents during later stages (3-13 mm) of oogenesis. Rickert's findings were fully confirmed by Marechal (1907) in an elegantly illustrated study of two sharks and two teleosts.

Within the context of the reproductive system, some other notable contributions should be mentioned. Woods (1902) carried out a quantitative study in which he discovered that the germ cells of the spiny dogfish shark Squalus acanthias originate in endoderm associated with the yolk sac and migrate apparently by amoeboid movement to the germinal ridge of the developing gonad. Borcea's (1905) classical study of the urogenital system contains the definitive statement on the structure of the nidamental (=shell) gland of the oviduct and its role in the secretion and morphogenesis of the egg case.

By the end of the nineteenth century, the descriptive aspects of chondrichthyan development were well known. The state of knowledge compared favorably with what was available for the more intensively studied domestic fowl and selected amphibians. Building on this foundation, future investigators would be able to formulate a series of questions based upon a functional and analytical approach to the study of development.

Comparative Embryology

Comparative studies of fish development constitute one of the glories of nineteenth century embryology. The comparative method as a formal concept in morphology has its roots in German Naturphilosophie and was extended to embryology by Dollinger and von Baer. Subsequently, Johannes Muller and others incorporated it into physiology. Already well established, comparative embryology was thrust into a position of prominence with the publication of Darwin's On the Origin of Species. Darwin stated that much could be learned about evolution through embryological studies. Fritz Mller (1864) and Haeckel (1866) considered that embryology, especially comparative embryology, was paramount in understanding evolution, i.e., "ontogeny recapitulates phylogeny". The new evolutionary impetus combined with the pre-existing research agenda drove comparative studies of fish development ahead at a rapid clip. Fishes provided a intellectual feast for evolutionary comparative embryologists.

The four major groups of fishes occupy key positions in vertebrate evolution. Agnaths (hagfish, lampreys) are the most primitive representatives of the craniate-vertebrate lineage. Chondrichthyans (sharks, rays, chimaeras) are among the oldest extant jawed vertebrates and incorporate the definitive vertebrate Bauplan and developmental program. Sarcopterygians comprise lungfishes, coelacanths, and tetrapods. The Actinopterygian or ray-finned fishes (Polypterus, sturgeons, gars, Amia, and teleosts) represent a group of bony fishes that evolved in a lineage that culminated in the teleosts, one of the most successful groups of vertebrates. Not only was there evolution at the phyletic level, but the process of development itself underwent evolutionary change.

Once embarked upon the comparative venture, the availability of study material became a problem. Many fishes of interest were found only in the fresh waters of the tropics, and some, such as the Australian lungfish Neoceratodus, were only discovered in the last quarter of the nineteenth century. Coincident with the rise of interest in comparative embryology, geographical regions that harboured fish species of interest became more accessible, the fishes themselves became better known, and the technology of travel improved. Even so, studying the embryology of fishes under primitive field conditions was a major task. Individuals such as Budgett, Kerr, and Semon are part of the golden age of exploration.

The hagfishes and lampreys are the most primitive representatives of the craniatevertebrate lineage. Both are jawless. Hagfishes are common, marine, bottom-dwelling fish in cold, inshore waters and in deeper waters off continental shelves. Recently, hagfishes have been considered to be the sister group to all remaining vertebrates. Virtually everything that is known about their embryonic development is derived from collections of Eptatretus (=Bdellostoma) stout eggs obtained by Bashford Dean and Franz Doflein in Monterey Bay, California, on the Pacific coast of North America in the 1890's. For all intents and purposes, there have been no modern embryological studies of hagfish except for Gorbman (1983), Gorbman and Tamarin (1986), and Wicht and Northcutt (1992), who used Dean's material. For this discussion, research on the Atlantic hagfish Myxine will not be considered because there is so little of it. Doflein (1898) and Dean (1898, 1899) described embryonic development.

Dean's 1899 extraordinary monograph gives a definitive account, sumptuously illustrated with colour lithographs, of hagfish development from early cleavage through hatching. Unfortunately, because the monograph is not readily accessible, this research has been overlooked. Most of Dean's account is based on the observation of whole mounts. Sections of blastula and gastrula stages are illustrated. The eggs are large ellipsoids about 22 mm in length by 8 mm in width. They are heavily yolked, and there is a germinal disc at the animal pole. Cleavage is meroblastic.

The process of embryo formation coincides with the epibolic movement of the blastoderm over the yolk mass. The equivalent of a germ ring is present and the embryo is formed along an anterior-posterior axis as the blastoderm moves over the yolk mass. The only section of a gastrula reveals the absence of bottle cells, although this could be a function of the plane of section. Both a cellular "periblast" and an underlying yolk syncytium are depicted. The neural tube is formed by the closure of the medullary folds. Much information is given on organogenesis and the later phases of embryonic development. Additional descriptions of early cleavage stages are given by Gudger and Smith (1931). A description of development of the head, brain and associated organs based on the analysis of sectioned material was published by von Kupffer (1899).

Lampreys are widely distributed throughout the waters of the northern and southern temperate zones. They were known to zoologists such as Home, Rathke, and Miller who worked on aspects of their reproductive anatomy. It is surprising, then, that embryological research was only undertaken in the mid-nineteenth century and that comprehensive studies only appeared in the latter part of the nineteenth and in the early twentieth centuries. The first substantive description of lamprey development was that of Schultze (1856), who amongst other observations demonstrated the somewhat unequal holoblastic cleavage of the egg. Aspects of gastrulation were described by Balfour (1880), Goette (1890), and Hatta (1907). Both Goette and Hatta demonstrated the presence of bottle cells and involution of surface cells through the blastopore. Other contributions of note are those of Scott (1887), von Kupffer (1890), and Selys-Longchamps (1910). Older literature on lamprey development is listed in Ziegler (1902). Piavis (1971) provides a modern account of development and a list of the more recent literature.

Two main lines of evolution are recognized within the class Chondrichthyes: the holocephalans and elasmobranchs. When chondrichthyan development was discussed previously, only the elasmobranchs (sharks, rays, and skates) were considered. Now the holocephalans (chimaeras or ratfishes) will be considered. Chimaeras are oviparous, bottom-dwelling marine fishes usually found in the deep waters off the continental shelf, except in those regions where the inshore waters are cold, e.g., the northwest coast of North America and New Zealand. Because of their deep water habitat, it has been difficult to procure their eggs for study.

The major sources of information on chimaeroid development are Schauinsland's (1903) and Dean's (1906) monographs. Schauinsland studied Callorhinchus mil, whereas Dean's study was mostly based on the eggs and embryos of Hydrolagus (=Chimaera) colliei collected in Monterey Bay, California. Dean's monograph is remarkably complete. Dean begins with an account of the breeding habitats of Hydrolagus and egg case formation and then describes in detail development from fertilization through hatching. Chimaeras are oviparous. Their eggs, moderately large, oblate spheroids (25 x 20 mm), are enclosed in an elaborate egg case composed of structural proteins that differ considerably from those of elasmobranchs. Cleavage and subsequent development is meroblastic and closely resembles that of sharks. A closely spaced series of stages from early cleavage through the mid-phases of development is illustrated with surface views of the blastoderm and drawings of the embryos.

This series is supplemented with many illustrations of sectioned material, especially from the blastula and gastrula stages. One developmental characteristic that sets chimaeras apart from elasmobranchs is the pronounced elongation of the post-anal region of the body, which begins at the tailbud stage and soon gives the young embryo an almost eel-like appearance. The studies of Schauinsland and Dean remained unchallenged for many years. Only in the late twentieth century has interest in chimaeroid development revived.

The osteichthyans or bony fishes include the sarcopterygians and actinopterygians. The sarcopterygians (fleshy-finned fishes) are now considered to include the tetrapods, two extinct groups of fishes, and two extant groups of fishes: coelacanths and dipnoans (lungfishes). These fishes are of considerable interest because the "lungfishes,coelacanths, or both together are the losest living relatives of tetrapods".

The living coelacanth Latimeria chalumnae is the sole surviving representative of a once common group. The only known population occurs in the Comoros Islands off the east coast of Africa. No embryological research was carried out on Latimeria during the nineteenth century because it was not discovered until 1938. However, in 1975 Smith et al. discovered that Latimeria is viviparous when they found five nearterm pups in the oviduct.

Up to this time only two gravid females have been collected. Wourms et al. (1991) described aspects of viviparity and the maternal-embryonic relationship using the embryos and maternal tissues from the Smith et al.

(1975) specimen. This material was also used by Bemis and Northcutt (1991) in their study of the innervation of cranial muscles and Northcutt and Bemis (1993) in their study of cranial nerves. Fricke and Frahm (1992) described full-term pups obtained from the second gravid female. Coelacanth eggs are enormous, up to 9 cm in diameter and with a dry weight of 184 g. They are heavily yolked and contain yolk platelets. Although no information is available on early embryonic development, it is reasonable to assume that development is meroblastic. Because the coelacanth is on the verge of extinction, it is highly unlikely that any substantial body of embryological knowledge will be forthcoming.

The dipnoans or lungfishes comprise the other extant group of sarcopterygian fishes. There are three genera of living lungfishes: Protopterus in Africa, Lepidosiren in South America, and Neoceratodus in Australia. Lepidosiren and Protopterus were discovered nearly simultaneously in the late 1830's and Neoceratodus in 1870. Lungfishes as "intermediate forms" readily fitted into the emerging evolutionary paradigm of the nineteenth century. Thus, the study of lungfish development became one of the most sought after goals of comparative embryology. It entailed expeditions to three continents that involved considerable risk and in some instances loss of life, viz. Budgett.

The major accounts of lungfish development appeared during a brief sixteen-year span at the end of the nineteenth and in the early twentieth centuries. The embryology of the lungfish Neoceratodus was first described by Semon (1893) in a series of 47 stages that extend from fertilization through post-hatching juveniles. This report, which was based on the description of external features, was complemented by a study of gastrulation based on the analysis of sectioned material. Much of the previous research was summarized and new aterial was incorporated into a publication that was part of Keibel's series of normal tables of development and others hae re-investigated the development of Neoceratodus.

In much the same fashion, Kerr (1900) published the first account of the development of the South American lungfish, Lepidosiren. An account of external development of closely spaced states from fertilization through post-hatching juveniles was given. Unlike the Australian lungfish, embryos f Lepidosiren develop external gill filaments before hatching and retain them for at least40 days. A second part of Kerr's account included a comparison with the development of the African lungfish Protopterus. The first account of its development was given by udgett (1901). This preliminary report contained information on the reproduction and development of other African fishes, most notably Polypterus. Budgett died in 1904 from malaria contracted in Afica. Kerr, Assheton, and thers acted as Budgett's scientific executors, ensuring the publication of his research. The definitive report on the development of

Protopterus was made by Kerr in the same volume of the Keibel series that contained the normal table of Lepidosiren development (Kerr, 1909).

Thus far, little has been said about lungfish development. Eggs are approximately 3 mm in diameter with a moderate amount of yolk. Cleavage is holoblastic and unequal. A blastopore develops in a subequatorial position. Bottle cells are present. Surface cells involute through the blastopore. The neurula is characterized by a flat neural plate that folds to form the neural tube. In general, lungfish development, especially that of the Australian lungfish, bears an uncanny resemblance to that of urodele amphibians.

Caution, however, is advised because of the need for a fine-grained analysis of the early (blastula-neurula) phases of development using contemporary methods. The actinopterygian or ray-finned fishes are of particular interest to comparative embryologists. They constitute a well-defined evolutionary lineage that culminated in the teleosts, one of the most successful, diverse, and species-rich groups of vertebrates. Based on ther evolutionary relationships, they provide a unique opportunity for investigating the transition from the primitive, holoblastic pattern of fish development to the specialized meroblastic pattern characteristic of teleosts. The actinopterygian fishes comprise three subclasses: Cladistia, Chondrostei, and Neopteryg.

The Cladistia are the most basal group of the surviving cladistians are represented by two genera, Polypterus and Calamoichthys, which are in the order Polypteriformes. In some classifications, e.g., Nelson (1994), the Polypteriformes are placed in the subclass Chondrostei together with the sturgeons. Both Polypterus, the sturgeons and paddlefish are considered to be part of the lineage that led to the teleosts. Polypterus, the bichir, and Calamoichthys are from Africa. Little or nothing is known about the development of Calamoichthys.

The situation is somewhat better with respect to Polypterus. The first, very fragmentary report of its development was made by Budgett (1901). Following Budgett's death in 1904, Kerr worked up Budgett's collections and notes and published the only extensive account of the development of Polypterus. Almost all aspects of development from fertilization through hatching and also post-hatching larvae are described in 36 stages. Eggs are small, slightly flattened spheres (0.9 x 1.3 mm along the major and minor axes), moderately yolked, and contain yolk platelets. Cleavage is holoblastic and unequal. Unfortunately, information on the critical events of gastrulation is scanty. At the onset of gastrulation, what appears to be a very wide blastopore occupies an equatorial position. Subsequently, the lateral margins of the blastopore extend and meet to form a circle nearly as large as the circumference of the egg. This marks the advancing boundary of the surface cells of the animal pole during epiboly. Epiboly is pronounced and initially a very large yolk plug is formed. Subsequently, it is reduced in size. The marginal quality of the single section of a gastrula stage illustrated by Kerr (1907) does not allow comment on the presence or absence of bottle cells. Better sections through early neural plate stages confirm the presence of a blastopore that is continuous with a large archenteron. During neurulation, a flat neural plate folds to form a neural tube.

The embryo passes through typical tail bud stages and eventually hatches as a larva with external gill filaments. According to Arnoult (1964), the rate of development is quite rapid, i.e., 48 hours from fertilization to hatching. Although there are differences, the development of Polypterus resembles that of the sturgeon. Obviously, Polypterus is a prime candidate for further study. The prospects are excellent for it is in the aquarium trade and can be spawned in captivity. In point of fact, Bartsch and Britz (1996) have just published a popular account of spawning and development that includes scanning electron micrographs of the neural plate stages. Some of their material was used in Piotrowski and Northcutt's (1996) study of cranial nerves.

The subclass Chondrostei has one extant order, the Acipenseriformes, that contains the sturgeons and paddlefishes. They are considered an offshoot of the lineage that led to the teleosts. Sturgeons occur in North America and Eruasia. Paddlefish are restricted to the United States and China. The development of sturgeons was first outlined by Kowalewsky et al. (1870) for the European sterlet (Acipenser ruthenus). A more detailed account of early development and embryogenesis was given by Salensky (1878, 1881). Somewhat later, similar studies were carried out on the eggs of the North American Atlantic sturgeon A. oxyrhinchus.

Recently, Ginsburg and Dettlaff (1991) and Dettlaff et aL (1993) have presented a detailed contemporary account of the development of the Russian sturgeon A. gueldenstaedt from fertilization through hatching. Eggs are 3.0-3.5 mm in diameter. Cleavage is holoblastic. A dorsal lip develops in an equatorial position as the apices of bottle cells contract. Involution of surface cells takes place through the blastopore. The neurula stage is characterized by a flat neural plate that subsequently folds to form the neural tube. According to Bolker (1993a), gastrulation is similar to that of the amphibian Xenopus. Morphogenetic movements have been analysed by Ballard and Ginsburg (1980). The mechanisms of gastrulation and mesoderm morphogenesis have been investigated by Bolker (1993a, 1993b). The embryology of the paddlefish (Polydon) did not become known until the 1960s. The early-and mid-phases of paddlefish development closely resemble that of sturgeons. During the later phases of development, embryos develop their characteristic elongated rostrum. A scanning electron microscopy study by Bemis and Grande (1992) has substantially extended the description of Polyodon development.

The subclass Neopteryg contains three divisions, viz. Ginglymodi (gars), Halecomorphi (Amia), and Teleostei. Gars (Lepisosteus) occur in eastern North America and Central America. The first account of the development of gars was by Alexander Agassiz (1878), a report that dealt exclusively with late embryonic and post-hatching stages. Using material provided by Agassiz, Balfour and Parker (1882) gave an account, albeit fragmentary, of some of the earlier phases of development. Most of their report, however, contains the first detailed account of embryogenesis. The egg is about 3 mm in diameter and heavily yolked. Cleavage is meroblastic. Sections of embryos reveal that the spinal cord originally is a solid rod that secondarily becomes hollow. Dean (1895) provided critical information on the early development of Lepisosteus from fertilization through gastrulation and compared its development with that of the sturgeon. Dean was the first to provide a timed series of developmental stages for both fishes. In addition to providing surface views of the external features of early development, Dean also based his descriptions on sections of a developmental series. Eycleshymer (1899) provided additional information on the cleavage process.

Recently, Long and Wourms (1991) examined early development with scanning electron microscopy. From these studies, the following view of early development of Lepisosteus emerges. Meroblastic cleavage produces a blastoderm atop a large, yolk sphere in which abortive cleavage furrows are present. Further studies are needed to ascertain whether the yolk mass undergoes any division. Epiboly carries the blastoderm margin toward the vegetal pole preceded by microvilli-bearing zone of the yolk surface similar to the yolk syncytial layer of teleosts. As epiboly of the blastoderm reaches the equator, a blastopore begins to form at one point along the rim of the blastoderm. Bottle cells are associated with the dorsal lip of the blastopore. By this time, cells of blastoderm are organized into an embryonic shield and a wide germ ring. As epiboly proceeds to completion, the germ ring narrows, the embryonic shield elongates, and a shallow, neural groove forms similar to that of teleosts.

Amia calva, the bowfin, is the sole extant representative of the order Amformes. It is confined to the freshwaters of eastern North America. Its development was made known in a flurry of papers that appeared during 1896-1906. Dean (1896a) reported on early development, i.e., fertilization through gastrulation. External views of successive stages were supplemented with sections of the corresponding stages.

The later stages of embryonic development as well as the development of the post-hatching larvae were reported on separately. Almost simultaneously, Whitman and Eycleshymer (1897) reported on reproduction and the cleavage stages of development. Again, external views of the whole egg were supplemented with sections of the corresponding stages.

It is interesting to note that Whitman raised the issues of both priority and unfair rivalry with Dean because Dean obtained eggs of Amia from the same small lake in Wisconsin in which Whitman had previously discovered spawning colonies. Ballard (1986a,b) has published a series of stages and rates of normal development and has investigated morphogenetic movements. He has also produced a provisional fate of development in Amia. Superficially, the development of Amia resembles that of Lepisosteus but with several important differences. The egg is an elongated sphere (2.2 X 2.8 mm along the major and minor axes). A germinal disc occupies about one-third of the egg volume. Cleavage is holoblastic but quite unusual. The yolk mass divides slowly and with apparent difficulty whereas the germinal disc is rapidly divided into many small cells.

In late cleavage, a blastoderm surrounds a yolk mass divided into giant blastomeres. Cells of the blastoderm can be distinguished as enveloping layer cells, epiblast, or hypoblast. As epiboly commences, morphogenetic movements within the blastoderm lead to the formation of the germ ring and embryonic shield. The early phases of embryo formation take place during epiboly of the blastoderm. One of the most significant differences between the development of Amia and Lepisosteus is that Amia lacks both a blastopore and bottle cells. It is disconcerting that Lanzi (1909) illustrated sections of a groove at the leading edge of the blastoderm that contains cells that appear to be bottle cells. For obvious reasons, ultrastructural studies are in order. Collazo et al. (1994) discuss in detail the phylogenetic implications of these comparative embryological studies of primitive actinopterygian fishes with respect to teleost gastrulation.

Teleosts constitute the remaining major group of actinopterygian fishes. The history of teleostean embryology has been discussed. The main emphasis of early studies was placed on defining the meroblastic pattern of teleostean development and piscine features of embryogenesis and organ and tissue formation. Some attention was given to the diverse embryonic adaptations for viviparity in different taxa. By the end of the nineteenth century, it was aparent that the early development of most teleosts was very similar but that mid-and late-phases diverged, yielding diverse embryonic morphology and larval forms adapted to specific life styles and habitats. Thus, the later phases of embryonic development, larval development, and life histories became the subjects of comparative studies of teleostean development during the late nineteenth century and the twentieth century.

Several factors favoured the growth of a comparative approach, especially the advent and rise of marine biology, biological oceanography, and fisheries biology. Fishes are the dominant group of marine vertebrates and are commercially important. Thus, the study of life histories and the embryonic and larval stages of inshore, pelagic, and deep-sea fishes became a research agenda for the marine sciences. The results were often quite spectacular, e.g., the leptocephalus larvae of eels and the bizarre larvae of the stomiatoid fish Idiacanthus with its stalked eyes and trailing gut.

Studies of this type were pioneered by Alexander Agassiz (1882) and Ryder (1884a) in the United States, McIntosh and Prince (1890) in Great Britain, and Ehrenbaum (1909) in Germany, amongst others. Similar activities were undertaken at limnological stations and in freshwater fisheries biology. Popularization of the aquarium hobby and the successful reproduction in captivity of freshwater fishes imported from Africa, South America, and Asia such as the zebrafish, Danio rerio, established another source of material for comparative studies. Given their numbers and diversity, a wealth of information, mostly of a descriptive nature, was produced on the development of a great many species of teleosts.

Increased activity in the study of teleostean development was accompanied by the elaboration of several concepts and theoretical approaches which both aided and impeded further progress. In 1877, Haeckel set forth his Gastrea theory. Based on his observations of teleostean development, he applied the concept to both teleosts and chondrichthyans. Much of the literature on fish development from his time to the present has been concerned with the morphogenetic movements and embryo-forming mechanisms that are associated with gastrulation, a literature too voluminous to review. One major problem that arose from overzealous and inept efforts to use embryological studies to support evolutionary theory was the attempt to homologize the events and processes of gastrulation in teleosts and chondrichthyans with amphibians and amniotes. In this respect, it was Goette (1873) who first proposed that involution, i.e., flowing of a sheet of cells over an inpocketing such as a blastopore, has a prominent role in teleost gastrulatio.

At the risk of oversimplification, many of these studies were comparative in name only. There was little or no conceptual basis for interpreting the mass of information. Attempts, such as that of Schultz (1910), to establish a well-defined conceptual basis for comparative embryological research were overlooked. (In passing, Schultz's 1910 monograph appears to be a direct precursor of de Beer's 1930 study of embryology and evolution.) Although the body of evolutionary theory was available, the influence of Haeckel's biogenetic law interfered with the productive application of it to many developmental issues. A growing anti-evolutionary bias tended to discourage this type of inquiry. Studies of individual species were often exceedingly well done, and the development of a single organ system was often compared in several different taxa and interpreted in the context of the evolutionary paradigm. Examples of the latter include Ryder's (1886c) study of heterocercy and the evolution of the fins and fin rays of fishes and Eigenmann's (1903) study of the abortive development of the eye in blind cave fishes. Eigenmann compared the development and degeneration of the eye in the blind cave fish Amblyopsis with that of the surfperch Cymatograster, a fish with a well-developed eye. In Amblyopsis, development of the eye begins normally, but then the lens degenerates and the entire eye regresses. Recently, Jeffery (1996), using the embryos of another blind cave fish Astyanax, has shown that programmed cell death of the lens eliminates lens induction from the cascade of inductive interactions required for normal eye development.

In general, however, the comparative method was not used to explore the evolution of development and the factors that brought about evolutionary change. It was not until the maturation of evolutionary thought in the twentieth century that a new set of concepts and approaches were produced and used to exploit the potential of comparative studies. Paramount among these are: (1) revived interest in the evolution of development and the role of development in evolution; (2) application of evolutionary theory to life histories; and (3) attempts to order the bewildering diversity of fish reproduction in terms of systematics or reproductive guilds.

Analysis of Development

The term "analysis of development" comprises a set of conceptual and methodological approaches that are used to study development, for example, functional morphology, biomechanics, physiology, biochemistry, and cellular biology. These approaches possess a strong experimental and causal bias. It is difficult to say when such analytical approaches to the study of development were introduced.

Although its origins are obscure, the experimental-analytical approach certainly was recognizable in the eighteenth century studies of Spallanzani (1729-1799) on insemination, Trembley (1710-1784) on hydra regeneration, and Broussonet (1789) on regeneration in fishes. By the nineteenth century, experimental physiology under the aegis of Claude Bernard and Johannes Muller had become a formalized science. Muller was a pivotal figure for he was both a physiologist and an embryologist.

It is probably through him and his many students that an analytical-experimental approach was transferred from physiology to development and became established in the mid-nineteenth century. For example, Haeckel (1869), a student of Mller, divided the larvae of the siphonophore Crystallodes in two and obtained smaller, but perfect larvae. It was only later in the nineteenth century that a more narrowly defined, conceptual approach emerged in the Entwicklungsmechanik of Roux and the mechanistic views of Loeb. Here I will discuss a few examples that demonstrate the analytical approach to fish development. The introduction of new embryological techniques opened up new areas of inquiry especially at the cellular level, and led to a series of important observations, a few of which will be discussed here. The organization of the teleostean egg, which differs considerably from that of most other vertebrates, offers unique opportunities for the study of fertilization.

The mature ovum is invested with an extracellular matrix, the egg envelope. At the animal pole of the egg, there is a funnel-shaped channel through the egg envelope termed the micropyle. Fertilization is accomplished by the passage of a single spermatozoan through the micropyle and its subsequent fusion with the ooplasm of the germinal disc. The micropyle was first described by von Baer (1835) and reinvestigated by Doyere in 1849, who appreciated its significance in fertilization but did not have any experimental evidence for its role.

Ransom (1867), in an extraordinary paper on the eggs of teleostean fishes, not only experimentally demonstrated fertilization, but also conducted experimental and analytical studies of egg organization and post-fertilization ooplasmic contractility. In a series of experiments, in which he was able to restrict or facilitate access of spermatozoa to the micropyle, Ransom (1867) observed the passage of a single spermatozoan through the micropyle and the subsequent events of the fertilization reaction, namely, changes in structure of the micropyle, disappearance (discharge) of the cortical "granules," the appearance of the perivitelline space, and the movement of cytoplasm into the germinal disc. He did not document the cytological details of spermegg fusion.

This was done later by Hoffmann (1881) who depicted aster formation, pronuclear fusion, and mitosis. Ransom (1866, 1867) also carried out experimental studies to ascertain the physical an chemical properties and structural organization of maturing oocytes, mature ovum, and fertilized egg. He consciously approached the egg from a cellular point of view. The eggs of some fishes, especially the stickleback Gasterosteus and the pike Esox, exhibit strong rhythmic contractions of the yolk mass shortly after fertilization and throughout the early cleavage stages. Contractile waves, originating at the animal pole, bring about transitory deformations of the yolk mass.

Ransom (1867) found that the contractions are independent of fertilization and appear to reside in what is now known as the yolk cytoplasmic layer. He conducted elaborate experiments to ascertain the "conditions" that regulate both contraction of the yolk mass and cytokinesis. The effects of a number o factors, viz. metabolic poisons, temperature, electricity, carbon dioxide, and oxygen were determined. He devised a system whereby he could subject eggs to electrical stimulation while simultaneously observing them through a microscope. Electrical stimulation excited peristaltic waves that were distinguishable from normal ones by their site of origin, rapidity of formation and progress, and greater deformation. Cleavage was not affected. Cold temperatures delayed the fertilization reaction and slowed cleavage whilst moderately warm temperatures accelerated cleavage. High temperatures irreversibly arrested contractions and cytokinesis. Cyanide had no effect on yolk contraction, but delayed cytokinesis. Morphine and chloroform inhibited contraction.

Carbon dioxide was found to be a potent inhibitor of both contraction and cytokinesis. On the basis of a series of experiments on oxygen requirements, he concluded: (1) oxygen at low levels is necessary for contraction and cytokinesis; (2) cytokinesis requires more oxygen than do yolk contractions; (3) embryonic stages involving cell multiplication and differentiation require much more oxygen than do early cleavage stages; and (4) the embryonic heart has a higher tolerance to oxygen deficiency than does striated trunk muscle. In a separate publication that also dealt with rhythmic yolk contractions, Ransom (1866) reported on the isolation of a protein "allied to myosin" in the yolk of all fishes, amphibia, and birds that he had examined. He did not, however, state that myosin is involved in yolk contraction. This apparent omission may be an instance of understatement.

In addition to his fertilization and contractility research, Ransom also described and isolated the germinal vesicle, the enlarged nucleus of early oocytes, and illustrated the multiple nucleoli within it. His pioneering efforts were not fully realized until the twentieth century. Most of the research during the latter half of the century involved histological studies of oocytes and eggs. In this respect, there were some interesting developments in the study of nuclear-cytoplasmic relationships and chromosomes. In a histological study of oogenesis, Van Bambeke (1893) depicts lampbrush chromosomes without recognizing them and reports on the "elimination of nuclear elements" from the nucleus into the ooplasm of early oocytes. It appears that he was describing ribosome production and storage. Lampbrush chromosomes were first described in the eggs of elasmobranchs (vide infra) and subsequently discovered in teleosts. The discovery of the behaviour of chromosomes during cell division was soon extended to include the cleavage stages of teleosts.

The ability to maintain and manipulate cells in vitro is an important tool of modern biology. Ross Harrison (1907) is usually credited with the "discovery" of cell-tissue culture. However, he had several forerunners, one of whom was Wilhelm His. His (1899) isolated single blastomeres and clumps of blastomeres from salmonid blastulae and maintained them in the transparent fluid portion of the egg yolk. Under these conditions, he was able to make longterm observations on the in vitro behaviour of blastomeres. He compared the behaviour of blastomeres from 0-4 day post-fertilization eggs (cleavage stages) with blastomeres from 5-12 day post-fertilization eggs (gastrula stages). He observed "amoeboid" movement and the extension and retraction of finger-like processes which contained clear cytoplasm (=hyaloplasm) in the 5-12 day blastomeres, but not in the 0-4 day blastomeres. In the later stage blastomeres, he also observed a transition in organization from single cells with finger-like processes, loosely arrayed in clusters, to round cells that are joined together to form epithelial sheets.

He compared amoeboid movement and contractility of blastomeres to muscle contraction. The cytoplasmic organization of living and fixed blastomeres was shown to be similar. The leading edge (lobopodia) of the blastomere consists of hyaline cytoplasm which may contain elements of a sparse fibrous reticulum (ER, cytoskeleton?) whereas the perinuclear cytoplasm is dense and contains both granular elements and a fibrous reticulum. Dynamic changes in the regional organization of the cytoplasm in living cells was noted. These observations were illustrated with both drawings and photomicrographs of sectioned cells. The in vitro observations of His are remarkably similar to those carried out by Trinkaus on the blastomeres of Fundulus many years later.

The first attempt to analyse the mechanism of embryo formation in teleosts was undertaken by Lereboullet in a set of descriptive and experimental studies of teratology. He used different temperature regimes and chemicals to induce abnormal development, and also tested the effect of pressure on development. Although he used over 200,000 eggs in his experiments, the results were inconclusive because the incidence of induced abnormalities in treated eggs was no greater than that in untreated eggs.

Oppenheimer (1936) speculated that Lereboullet worked with too many eggs and should have monitored each egg. Careful individual analysis of the many "spoiled eggs" in the experiments would have produced more conclusive results. Lereboullet's descriptive stdies were more fruitful. By tracing the progress of developmental anomalies, especially duplicated or partially duplicated embryos from their first visible origin, Lereboullet obtained insight into both their development and the normal process of embryogenesis. These studies caused him to focus on the embryo-forming dynamics of the germ ring. Eventually, he concluded that embryo formation in teleosts took place by the process that His (1874) later called concrescence.

The pioneering efforts of Ransom and Lereboullet were capitalized on by Loeb, Stockard, and others. They used a variety of physical and chemical parameters either to determine the physiological conditions required for normal development r to investigate the mechaisms of developmental regulation by perturbing those conditions. Loeb began his work on fish embryos in the 1890's primarily using the cyprinodontid Fundulus, found in abundance at Woods Hole. In a series of early experiments, he found that embryos, the hearts of which had been immobilized by potassium chloride, developed structurally norml circulatory systems but ones lacking blood circulation.

Chromatophores that normally migrated onto the surface of the vitelline vessels of the yolk sac failed to do so. Loeb suggested that migration would only occur if there was blood circulating in the vessels. Later, used his research on the effects of ions on embryos of Fundulus as the basis for his theory of physiologically balanced salt solutions. Stockard (1906, 1907) studied the effects of magnesium and lithium ions on the development of Fundulus in order to compare them with the development-perturbing effects already well known in sea urchins and frogs. At high ionic concentrations, cyclopean monsters with a single median eye were produce. By varying ion concentrations, stages intermediate between normal and cyclopean could be produced. The ionic perturbation interfered with the separation of the early, single eye-forming field into the right and left eye anlagen.

Although radioactivity was only discovered in the latter years of the nineteenth century, radioisotopes were soon employed in embryological research. For example, Tur (1896) studied the deleterious effects of radium radiation on catshark Scyliorhinus development, and Oppermann (1913) described a variety of developmental abnormalities that resulted from the fertilization of trout eggs with radium-irradiated spermatozoa. Fundulus has been a favourite experimental animal at Woods Hole and other marine laboratories on the east coast of the United States for over 100 years. Given its popularity, it is natural to inquire into its history. The first accounts of the development of Fundulus were given by Alexander Agassiz (1882) and Ryder (1886b).

Boyer (1892) was the first to specifically state that he had obtained eggs by stripping ripe females and artificially fertilizing them, and he also gave a more extensive account of the early stages of development. Who, then, introduced the eggs and embryos of Fundulus as experimental material and developed the methods for obtaining them? I speculate that John Ryder was the driving force behind Fundulus. Ryder was an embryologist's embryologist. In his brief life, he published numerous (215) papers, many of which dealt with fishes. From 1880 to 1886, he was in charge of embryological research at the U.S. Fish Commission Laboratory at Woods Hole. Much of his research on fish embryology required artificial fertilization of eggs.

The Fish Commission Laboratory was far better equipped for research than was the fledgling Marine Biological Laboratory. Consequently, many embryologists such as Boyer, Morgan, and others started their work there and continued as long as the cultural atmosphere was conducive to basic scientific research. Ryder left the Fish Commission Laboratory for a professorship at the University of Pennsylvania which he held until his death in 1895. Both Morgan and Loeb were on the faculty of Bryn Mawr and in close proximity to Ryder. Did their paths cross? In terms of what has come to be regarded as classical experimental embryology, the first experiments of consequence on embryonic fishes are based on the concrescence theory. A different approach to the study of fish embryology had been taken by His, who was more concerned with the events and mechanisms of development than their possible evolutionary significance. In his quest for a mechanistic theory, His (1874) proposed his "Konkreszenz Theorie" to explain the observed relationship between embryo formation, and the cellular material in the lateral arms of the germ ring.

As originally conceived by His, concrescence was a process by which the right and left lateral margins of the germ ring became apposed to one another in the forming embryo, so that as the two lateral margins progressively fused in a medial plane, the embryo grew backwards. In this theory, the embryo was formed by the union of the two halves of the germ ring so that successive regions along the anterior-posterior axis of the embryo were originally located in successively distal regions along the circumference of the germ ring. His applied his theory both to teleostean and chondrichthyan development. Although not implicitly stated, His's theory is a form of mosaic or determinate development. Either because it was challenging or because it was so succinctly stated, the concrescence theory soon was put to the test. The use of experiments to test a theory of development was a novel idea at that time and represents an early phase in the paradigm shift to experimental embryology.

The first person to test the concrescence theory was Kastschenko (1888), who operated on the embryos of the catshark Scyliorhinus. He carried out three sets of experiments but did not illustrate his results. In the first, he cut the germ ring on one side of a Stage 7, lancet-shaped embryo. A normal embryo developed. In another experiment, he destroyed the posterior end of the blastoderm but not the germ ring.

The anterior half of the embryo developed normally, but the posterior half did not. If His's theory were correct, the posterior half of the embryo should have developed because the tissue for its formation was intact in the germ ring. In the third experiment, the entire embryo at Stage 7 was separated at its two sides from the blastoderm, severing its connection to the germ ring. A normal embryo, with both sides of the body intact, continued to develop to the three somite or tailbud stage. According to His, a defective, partial embryo should have been produced because the cellular material needed to complete embryo formation remained in the germ ring.

Kastschenko's pioneer research was soon followed by more extensive studies by Morgan (1895) and Kopsch (1896, 1898), who tested theconcrescence theory in both chondrichthyans and teleosts. Kopsch's results differed somewhat from those of Kastschenko. Morgan's research is interesting not only because of its bearing on the concrescence theory, but also because of his experiments on early blastomeres. Morgan (1895) carried out two sets of experiments on the eggs of Fundulus. In the first set, he completely removed one of the first two blastomeres. The remaining blastomere divided, and a somewhat smaller blastoderm was produced. From it a single, normal embryo formed, albeit somewhat smaller than usual. In the second set, Morgan cut the germ ring on one side of an early embryo. The embryo continued to develop normally and underwent posterior elongation even though the germ ring remained intact only on the unoperated side.

The results contradict His's theory. In retrospect, His's concrescence theory was an excessively rigid statement of the process of confluence of germ-ring cells and their incorporation into the embryo. Before leaving the subject of concrescence, I wish to point out that His restated the theory in 1894, but, more importantly, in that paper he introduced physical models as a means of illustrating the mechanical basis of morphogenesis. Other approaches to the study of developmental mechanisms were undertaken during this period. Following his pioneering application of cell lineage analysis to the development of Clepsine, Whitman attempted to take a similar approach to teleosts. Both his attempts and those of Wilson (1891) were premature and doomed to failure because there was no sure way of identifying individual blastomeres and their progeny and tracing their fate during development.

The study of the physiology of embryonic nutrition is the last topic to be considered. This aspect of piscine development had its origin in the study of viviparous fishes. In many such fishes, there is a marked difference between the small size of the egg and the large size of the full-term embryo. How is this accomplished? The problem of embryonic nutrition in viviparous fishes was recognized as early as 1624 when Schoenveld postulated that embryos of Zoarces ingested ovarian fluid as a source of nutrients.

It was not until the nineteenth century that physiological and chemical studies of embryonic nutrition were attempted. Davy (1834) appears to be the first to have undertaken this line of investigation. He compared the mean weight of the egg (182 grains) of the electric ray Torpedo with that of the full-term embryo 479 grains). On the basis of the increase in weight, he concluded that the embryo received additional maternal nutrients during gestation, a correct conclusion but based on faulty data collection. Modern studies, reviewed in Wourms et al. (1988), show that there is actually a loss of organic weight during development of this fish. The apparent discrepancy arose because wet weight instead of dry or organic weight was measured. Teleosts also were studied. In 1867, Blake analysed the ovarian fluid of a surfperch and found that proteins and lipids were the major organic constituents and hence the source of nutrition. Analysis of ovarian fluid was next carried out on Zoarces by Stuhlmann (1887) and Kolster (1905), who found it to be protein-and lipid-rich. Interest in tis area of inquiry waned until it was revived by Ranzi (1932) and attracted the attention of Needham (1942).

3

DNA and Gene Expression

DNA Structure

Deoxyribonucleic acid or DNA, is a nucleic acid that contains the genetic instructions used in the development and functioning of all known living organisms (with the exception of RNA iruses). The main role of DNA molecules is the long-tem storage of information. DNA is often compared to a set of blueprints, like a recipe or a code, since it contains the instructions needed to construct other components of cells, such as protens and RNA molecules.

The DNA segments that carry this genetic information ae called genes, but other DNA sequences have structural purposes, or are involved in regulating the use o this genetic information. DNA consists of two long polymers of simple units called nucleotides, with backbones made of sugars and phosphate groups joined by ester bonds. These two strands run in opposite directions to each other and are therefore anti-parallel. Attached to each sugar is one of four types of molecules called bases.

It is the sequence of these four bases along the backbone that encodes information. This information is read using the genetic code, which specifies the sequence of the amino acids within proteins. The code is rad by copying stretches of DNA into the related nucleic acid RNA, in a process called tanscription. Within cells, DNA is organized into long structures called chromosomes. These chromosomes are duplicated before cells ivide, in a process called DNA replication. Eukaryotic organisms (animals, plants, fungi, and prtists) store most of their DNA inside the cel nucleus and some of their DNA in organelles, such as mitochondria or chloroplasts.In contrast, prkaryotes (bactera and archaea) store their DNA only in the cytoplasm. Within the chromosomes, chromatin proteins such as hitones compact and organize DNA. These compact structures guie the interactions between DNA and other proteins, helpingcontrol which parts o the DNA ar transcribed.

Properties

DNA is a long polymer made from repeating units called nucleotides.As first discovered by James D. Watson and Franis Crick, the structure of DNA of all species comprises two helical chains each coiled round the same axis, and each with a pitch of 34 Angstroms (3.4 nanometres) and a radius of 10 Angstroms (1.0 nanometres). When measured ina particular solution, the DNA chain measured 22 to 26 Angstroms wide (2.2 to 2.6 nanometres), and one nucleotide unit measured 3.3 A (0.33 nm) long. Although each individual reeating unit is very small, DNA polymers can be very large molecules containing millions of nucleotides. For instance, the largest human chromosome, chromosome number 1, is approximately 220 million base pair long.In living organisms, DNA does not usually exist as a single molecule, but instead as a pair of molecules that are held tightly together. These two long strands entwine like vines, in the shape of a double helix.

The nucleotide repeats contain both the segment of the backbone of the molecule, which holds the chain together, and a base, whih interacts with the other DNA strand in the helix. A base linked to a sugar is called a nucleoside and a base linked to a sugar and one or more phosphate groups is called a nucleotide. If multiple nucleotidesare linked together, as in DNA, this polymer is called a polynuclotide. The backbone of the DNA strand is made from alternating phosphate and sgar residues. The sugar in DNA is 2-deoxyribose, which is a pentose (five-carbon) sugar. The sugars are joined together by phosphate grups that form phosphodiester bonds between the third and fifth carbon atoms of adjacent sugar rings.

These asymmeric bonds mean a strand of DNA has a direction. In a double helix the direction of the nucleotides in one strand is opposite to their direction in the other strand: the strands are antiparalel. The asymmetric ends of DNA strands are called the 5? (five prime) and 3? (three prime) ends, with the 5' end having a terminal phosphate group and the 3' end a terminal hydroxyl group. Oe major difference between DNA and RNA is the sugar with the 2-deoxyribose in DNA being replaced by the alternative pentse sugar ribose in RNA.The DNA double helix is stabilized by hydrogen bonds between the bases attached to the two strans. The four bases found in DNA are adenine (abbreviated A), cytosine (C), guanine (G) and thymine (T).

These four bases are attache to the sugar/phosphate to form the complete nucleotide, as shown for adenosine monophosphate. These bases are classified into two types; adenine and guanine are fused five- and six-membered heterocyclic compounds called purines, while cytosine and thymine are six-membered rings called pyrimidines. A fifth pyrimidine base, called uracil (U),usually takes the place of thymine in RNA and differs from thymine by lacking a methyl group on its ring. Uracl is not usually found in DNA, occurring only as a breakdown product of cytosie. In addition to RNA and DNA, a large number of artificial nucleic acid analogues have also been created to study the proprieties of nucleic acids, or or use in biotechnology.

Grooves

Twi helical strands form the DNA backbone. Another double helix may be foundby tracing the spaces, or grooves, between the strands. These voids are adjacent to the base pairs and may provide a binding site. As the strands are not directl opposite ach other, the grooves are unequally sized. One groove, the major groove, is 22 Å wide and the other, the minor groove, is 12 Å wide. The narrowness of the minor groove means that the edges of the bases are more accessible in the major groove. As a result, proteins like transcription factors that can bind to specific seqences in double-stranded DNA usually make contacts to the sides of the bases exposed in the major groove.This situation varies inunusual conformations of DNA wthin the cel but the major and minor grooves are always named to reflect the differences in size hat would be seen if the DNA is twisted back nto the ordinary B form.

Base Pairing

Each type of base on one strand forms a bond with just one type of base on the other strand. Thi is called complementary base pairing. Here, purines form hydrogen bonds to pyrimidines, with A bonding only to T, and C bonding only to G. This arrangement of two nucleotides binding together across the double helix is called a base pair. As hydrogen bonds are not covalent, they can be broken and rejoined relatively esily. The two strands of DNA in a double helix can therefore be pulled apart like a zipper, either by a mechanical force or high temperature.As a result of this complementarity, all the information in the doube-stranded sequence of a DNA helix is duplicated on each trand, which is vital in DNA replication. Indeed, this reversible and specific interaction between complementary base pairs is criticalfor all the functions of DNA in living organisms.

The two types of base pairs form different numbers of hydrogen bonds, AT forming two hydrogen bonds, and GC formng three hydrogen bonds. DNA with high GC-cotent is more stable than DNA with low GC-content, but contrary to popular belief, this is not due to the extra hydrogen bond of a GC base pair but rather the contribution of stacking interactions (hydrogen bonding merely provides specificity of the pairing, not stability). As a result, it is both the percentage of GC base pairs an the overall length of a DNA double helix that determine the strength of the association btween the wo strands of DNA. Long DNA helices with a high GC content have stronger-interacting strands, while short helices with high AT content have weaker-interacting strands.In biology, parts of the DNA double helix that need to separate easily, such as the TATAAT Pribnow box in some promoters, tend to have a high AT content, making the strands easier to pull apart. In the laboratory, the strength of this interaction can be measured by finding the temperature required to break the hydrogen bonds, their melting temperature (also called Tm value). Whe all the base pairs in a DNA double helix melt, the strands separate and exist in solution as two entirely independent moleculs. These single-sranded DNA molecules (ssDNA) have no single common shape, but some conformations are ore stable than others.

Sense and Antisense

A DNA sequence is called "sense" if its sequence is the same as that of a messenger RNA copy that is translated into protein. The sequence on the opposite strand is called the "ntisense" sequence. Both sense and antisense sequences can exist on different pats of the same strand of DNA (*i.e.* oth strands contain both sense and antisense sequences). In both prokaryotes and eukaryotes, antisense RNA sequences are produced, but the functions of these RNAs are not entirely clear. One proposal is that antisense RNAs are involved in regulating gene epression through RN-RNA base pairing.A few DNA sequences in prokaryotes and eukaryotes, and more in plasmids and viruses, blu the distinction between sense and antisense strands by having overlapping gnes. In these cases, some DNA sequences do double duty, encoding one protein when read along one strand, and a second protein when read in the opposite direction along the other strand. In bacteria, this overlap may be involved in the regulation of gene transcription, while in viruses, overlapping genes increase the amount of information that can be encoded within the small vral genome.

Supercoiling

DNA can be twisted likea rope in a process caled DNA supercoiling. With DNA in its “relaxed” state, a strand usually circles the axis of the double helix once every 0.4 base pairs, but if the DNA is twisted the strands become more tightly ormore loosely wound. If the DNA is twisted in the direction of the helix, this is positive supercoiling, and the bases are held more tightly together. If they are twisted in the opposite direction, this is negative supercoiling, and the bases come apart mre easily. In nature, most DNA has slight negative supercoilng that is introduced by enzymes called topoisomerases. These enzymes are also needed to relieve the twistin stresses introduced into DNA strands during processessuch as transcription and A replication.

Aternate DNA Structures

DNA exists in many possible coformatins that incude A-DNA, B-DNA, and Z-DA forms, athough, only B-DNA and Z-DNA have been directly observed in functional organims. The conformation that DNA adopts depens on the hydration level, DNA sequence, the amount and direction of supercoiling, chemical modifications of the bases, the type and concentration of metal ions, as well as the presence of polyamines in solution. The fist published reports of A-DNA X-ray diffrction patterns—and also B-DNA used analses based on Patterson transforms that provided only a limited amount of structural informatin for oriented fibres of DNA. An alternate analysis was then proposed by Wilkins *et al.*, i 1953, for the in vivo B-DNA X-ray diffraction/scattering paterns of highly hydrated DNA fibres in terms of squares of Bessel functions.

In the same journal, James D. Watson and Francis Crick presented their molecularmodeling analysis of the DNA X-ray diffraction patterns to suggest that the structure was a doube-helix. Although the 'B-DNA form' is most common under the conditions found in cells, it is not a well-defined conformatio but a family of related DNA conformations that occur at the high hydration levels present in living cells.

Their corresponding X-ray diffraction and scattering patterns are characteristic of molecular paracrystals with a significant degree o disorder. ompared to B-DNA, the A-DNA form is a wider right-handed spiral, with a shallow, wide minor groove and a narrower, deeper major groove. The A form occurs under non-physiological conditions in partialy dehydrated samples of DNA, while in the cell it may be produed in hybrid pairings of DNA and RNA strads, as well as in enzyme-DA complexes. Segments of DNA where the bases have been chemically modified by methylation may undergo a larger change in conformation and adopt the Z form. Here, the strands turn about the helical axis in a left-handed spiral, the opposite of the more common B form. These unusual structures can berecognized by specific Z-DNA binding proteins and may be involved in the regulation oftranscription.

Alternate DNA Chemistry

For a number of years exobiologists have proposed the existence of a shadow biosphere, a postulated microbial biosphere of Earth that uses radically different biochemical and molecular processes than currently known life. One of the proposals was the existence of lifeforms that use arsenicinstead of phosphorus in DNA. A December 2010 NASA press conference revealed that the bacterium GFAJ-1, which has evolved in an arsenic-rich environment, is the first terrestrial lifeform found which may have this ability. The bacterium was found in Mono Lake, east of Yosemite National Park. GFAJ-1 is a rod-shaped extremophile bacterium in the family Halomonadaceae that, when starved of phosphorus, may be capable of incorporating the usually poisonos element arsenic in its DNA. This discovery lends weight to the long-standing idea that extraterrestrial life could have a different chemical makeup from life on Earth. The research was carried out by a team led by Felisa Wolfe-Simon, a geomicrobiologist and geobiochemist, a Postdoctoral Fellow of the NASA Astrobiology Institute with Arizona State University.

Quadruplex Structures

At the ends of the linear chromosomes ae specialized regions of DNA called telomeres. The main function of these regions is to allow the cell to replicate chromosome ends using the enzyme telomerase, as the enzyme that normally replicate DNA cannot copy the extreme 32 ends of chromosomes. These specialized chromosome cps also help protect theDNA ends, and stop the DNA repair systems in the cell from treating them as damage to be corrected.

In human cells, telomeres are usually lngths of single-stranded DNA containing several thousand repeats of a simple TTAGGG sequence. These guanine-rich sequences may stabilize chromosome ends by forming structures of stacked sets of four-base units, rather than the usual ase pairs found in other DNA molecules. Here, four guanine bases form a flat plate and these flat four-base units then stack on top of each other, to form a stable G-quadruplex structure. These structures are stabilized by hydrogen bonding between the edges of the bases and chelation of a metal ion in the centre of each four-base unit.

Other structures can also be formed, with the central set of four bases coming from either a single strand folded around the bases, or several different parallel strands, each contributing one base to the central structure. In addition to these stacked structures, telomeres also form large loop structures called telomere loops, or T-loops. ere, the single-stranded DNA curls around in a long circle stabilized by telomere-binding proteins.

At the very end of the T-loop, thesingle-stranded telomere DNA is held onto a egion of double-stranded DNA by the telomere strand disrpting thedouble-helical DNA and base pairing to one of the two strands. This triple-stranded structure is called a dislacement oop or Dloop.

Branched DNA

In DNA fraying occurs when non-complementary regions exist at the end of an otherwise complmentary doublestrand ofDNA. However, branched DNA can ocur if a third strand of DNA is introduced and contains adjoining regions able to hybridize with the frayed regions of the pre-existing double-strand. Althoug the simpest examle of branched DNA involvs only three strands of DNA, complexes involving additional strands and multiple braches are lso possible. Branched DNA can be used in nanotechnology to construct geometric shapes.

Chemical Modifications

Base Modifications

The expression of gnesis influenced byhow the DNA is packaged in chromosomes, in a stucture called chromatin. Base modifications can be involved in packaging, with regions that have low or no gene xpression usually containing high levels of methylation of cytosine bases. For example, cytosine methylation, produces 5-methylcytosine, which is important for X-chromosome inactivation.

The average leve of methylation varies between organisms—the worm Caenorhabditis elegans lacks cytosine methylation, while vertebrates have higher levels with up to 1% of their DNA containing 5-methylcytosine. Despite the importance of 5-methylcytosine, it can deaminate to leave a thymine base, so methylated cytosines are particularly pone to mutations. Other base modifications include adenine methylation in bacteria, the presence of 5-hydroxymethyl-cytosine in the brain, and the glycosylation of uracil to produce the "J-base" inkinetoplastids.

Damage

DNA can be damaged by many sorts of muagens, which change the DNA sequence. Mutagens include oxidizing agents, alkylating agents and also high-energy electromagnetic radiation such as ultraviolet lightand X-rays. The type of DNA damage produced depends on the type of mutagen. For examle, UV light can damage DNA by producing thymine dimers, which are cross-inks betweenpyrimidine bases.

On the other hand, oxidants such as free radicals or hydrogen peroxide produce multiple foms of damage, including base modifications, particularly of guanosine, and double-strand breaks. A typical human cell contains about 150,000 bases that have suffered oxidative damage. Of these oxidative lesions, the most dangerous are double-strand breaks, as these are difficult to repair and can produce point mutations, insertion and deletions from the DNA sequence, as well as chromosomal translocations. Many mutagens fit into the space between two adjacent base pairs, this is called intercalation. Most intercalators are aromatic and planar molecules; examples include ethidium bromide, daunomycin, and doxorubicin. In order for an intercalator to fit between base pairs, the bases must eparate, distorting the DNA strands by unwinding of the double helix.

This inhibit both transcription and DNA replication, causing toxicity andmutations. As a result, DNA intercalators are often carcinogens, and benzo[a]pyrene diol epoxide, acridines, aflatoxin and ethidium bromide are well-known examples. Nevertheless, due to heir ability to inhibit DNA transcription and replication, other similar toxins are also used in chemotherapy to inhibit rapidy growing cancer cells

Biological Functions

DNA usually occurs as linear chromosomes in eukaryotes, and circular chromosomes in prokaryotes. The set of chromosomes in a cell makes up its genome; the human genome has approximately3 billion base pairs of DNA arranged into 46 chromosomes.

The information carried by DNA is held in te sequence of pieces f DNA called genes. Transmission of genetic information in genes is ahieved via complementary base pairing. For example, in transcription, when a cell uses the iformation in a gene, the DNA sequence is copie into a complementary RNA sequence through theattraction between the DNA and the correct RNA nucleotides. Usually, this RNA copy is then used to make a matching protein sequence in a process called translation, which depends on the same interaction between RNA nucleotides. In alternative fashion, a cell may simply copy its genetic informaion in a process alled DNAreplication.

Gees and Genomes

Genomic DNA is tightly and orderly paced in the process called DNA condensation to fit the small available volumes ofthe cell. In eukaryotes, DNA is located in the cell nucleus, as well as small amounts in mitochondria and chloroplsts. In prokaryotes, the DNA is held within an irregularly shaped body in the cytoplasm called the nucleoid.

The genetic information in a genome is held within genes, and the complete set of this information in an organism is called its genotype. A gene is a unit of hredity and is a region of DNA that influences a particular characteristic in an organism. Genes contain an open reading frame that can be transcribed, as well as regulatory sequences such as promoters and enhancers, which control the transcription of the open reading frame.In many species, only a small fraction of the total sequence of the genome encodes protein.

For example, only about 1.5% of the human genome consists of protein-coding exns, with over 50% of human DNA consisting of non-coding repetitive sequences.The reasons for the preence of so much no-coding DNA in eukaryotic genomes and the extraordinary differences in genome size, or C-value, among species represent a long-standing puzzle known as the"C-value enigma". However, DNA sequences that do not code protein may still encode functional non-coding RNA molecules, which are involved in the regulation of geneexpression.Some non-coding DNA sequences play structural roles in chromosomes. Telomeres and centromeres typically contain few genes, but are important for the function and stability of chromosomes.

Anabundant form of non-coding DNA in humans are pseudogenes, which are copies of genes that have been disabled by mutation.These sequences are usually just molecular fossils, although they can occasionally serve as raw genetic material for the creation of new genes through the process of gene duplication and divergence.

Transcription and Transation

A gene is a sequence of DNA that contains enetic information and can influence the phenotype of an organism. Withina gene, he sequence of bases along a DNA strand defines a messenger RNA sequence, which then defines one or more protein sequences. The relationship between the nucleotide sequences of genes and the amino-acid sequences of proteins is determined by the rules of translation, known collectively as the genetic code. The genetic code consists of three-letter 'words' called codons formed from a sequence of three nucleotides (*e.g.* ACT, CAG, TTT).In transcription, the codons of a gene are copied into messenger RNA by RNA polymerase.

This RNA copy is then decoded by a ribosome that reads the RNA sequence by base-pairing the messenger RNA to transfer RNA, which carries amino acids. Since there are 4 bases in 3-letter combinations, there are 64 possible codons (4^3 combinations). These encode the twenty standard amino acids, giving most amino acids more than one possible codon. There are also three 'stop' or 'nonsense' codons signifying the end of the coding region; these are the TAA, TGA and TAG codons.

Replication

Cell division is essential for an organism to grow, but, when a cell ivides, it must replicate the DNA in its genome so that the two daughter cells have the same genetic information as their parent. Th double-stranded structure of DNA povides a simple mechanism for DNA replication. Here, the two strands are separated and thn each strand's complementary DNA sequence isrecreated by an enzyme called DNA polymerase. This enzyme makes the complementary strand by finding the corrct base through complementary base pairing, and bonding i onto the original strand. As DNApolymerases can only extend a DNA strand in a 5' to 3' direction, different mechanisms are used to copy the antiparallel strands of the double helix.In this way, the base on the old strand dictates which base appears on the new strand, and the cell endsup with a perfect copy of its DNA.

Interactions with Roteins

All the functions of DNA depend on interactions with proteins. These protein interactions can be non-specific, or the protein canbind specifically to a single DNA sequnce. Enzymes can also bind to DNA and of these,the polymerases that copy the DNA basesequence in transcription and DNA replicatin are particularly important.

DNA-Binding Proteins

Structural proteins that bind DNA are well-undertood examples of non-specific DNA-protein intections. Within chromosomes, DNA is held in complexes with structural protein. These proteins organize the DNA into a compact structure called chromatin. In eukayotes this structure involves DNA binding to a complex of small basic protein called histones, while in prokaryotes multiple types of proteins are involved. The histones form a disk-shaped complex called a nucleosome, which contains two compete turns of double-stranded DNA wrapped around its surface. These non-specific interactions are formed through basic residues in the histones making ionic bonds to the acidic suar-phosphate backbone of the DNA, and are therefore largely independent of the base sequence.

Chemical modifications of these basic amino acid residues include methylation, phosphorylation and acetylation. These chemical changes alter the strength o the interaction between the DNAand the histones, making the DNA more or less accessible to transcription factors and changing the rate of trancription. Other non-specific DNA-binding proteins in chromatin include the high-mobility group proteins, whch bind to bent or distorted DNA. These proteins are important in bending arrays of nucleosomes and arranging them into the larger structures that make up chomosomes.A distinct group of DNA-binding proteins are the DNA-binding proteins that specficaly bind single-stranded DNA. In humans, replication protein A is the best-understood member of this family and is used in processes where the double hlix is separated, including DNA relication, recombination and DNA repair.

These binding proteins seem o stabilize single-stranded DNA and protect it from forming stem-loops or being degraded by nucleases.In contrast, other proteins have eolved to bind to particular DNA sequences. The most intensively studied of these are the various transcription factors, which are proteins that regulate transcription. Each transcription factor bids to one particular set of DNA sequences and activates or inhibits the transcription of genes that have these sequences close to their promoters. The transcription factors do this in two ways. Firstly, they can bind the RNA polymerase responsible for transcription, either directly or through other mediator proteins; this locates the polymerase at the promoter and allows it to begin transcription. Alternatively, transcription factors can bind enzymes that modify the histones at the promoter; this will chnge the accessibility of the DNA templat to the polymerase.

As these DNA targets can occur throughout an organism's genome, changes in the activity of one type of transcription factor can affect thousands of genes.Consequently, these proteins are often the targets of the signal transduction processes that control responses to environmental changes or cellular differentiation and development. The specificity of these transcripton factors' interactions with DNA come from the proteins making multipl contacts to the edges of the DNA base, allowing them to "read" the DNA sequence. Most of these base-interctions are made in the major groove, where thebases are most accessble.

DNA-Modifying Enzymes

Nucleases and Ligases

Ncleases are enzymes that cut DNA strands by catalyzing the hydrolysis of thephosphodiester bonds. Nucleases that hydrolysenucleotides from the ends of DNA strands are called exonucleases, while endnucleases cut within strands. Te most frequently used nucleases in molecular biology are the restricion endonucleases, which cut DNA at specific sequences. For instance, the EcoRV enzyme shown to the left recognizes the 6-base sequence 52-GAT|ATC-32 and makes a cut at the vertical line. In nature, these enzymes protect bacteria against phage ifction by digesting the phage DNA when it enters the bacterial cell, acting as part of the restriction modification system. In technology, these sequence-specific nucleases are sed in molecular cloning and DNA fngerprinting. Enzymes called DNA ligses can rejoin cut or broken DNA strands. Ligases are particularl important in lagging strand DNA replication, as they join tgether the short segments of DNA produced at the replication for into a complete copy of the DNA teplate. They are also used in DNA repair and genetic recombination.

Topoisomerases and Helicases

Topoismerases are enzymes with both nuclease and ligase activity. These proteins change te amount of supercoiling in DNA. Some of theseenzymes work by cutting the DNA helix and allowing one section to rotate, thereby reducing its level of supercoilin; the enzyme then seals the DNA break. Other types of these enzyme are capable of cutting one DNA helix and the passing a second strand of DNA through this break, before rejoining the helix. Topoisomerases are required or many proceses involving DNA, such as DNA replication and transcription. Helicases are proteins that are a type of molecular motor. They use the chemical energy in nucleoside triphosphates, predominantly ATP, to break hydrogen bonds etween bases and unwind the DNA double helix into single strands. These enzymes are essential for most processes wher enzymes need to access the DNA bases.

Polymerases

Polymerases are enzymes that synthesize polynucleotide chains from nucleoside triphosphates. The sequence of their products are copies of existing polynucleotide chains - which are called templates. These enzymes function by adding nucleotides onto the 32 hydroxyl group of he previous nucleotide in a DNA strand. As a consequence, all polymerases work in a 52 o 32 direction. In the active site of these enzymes, the incoming nucleoside triphosphate base-pairs to the template: this allows polymerases to accurately synthesize the complementary strand of their template. Polymerases are classified just as to the type o template that theyuse. In DNA rpliction, a DNA-dependent DNA olymerase makes a copy of a DNA sequence.

Accuracy is vital in this process, so many of these polymerases have a proofreading activity. Here, the polymerase recognizes the occasional mistakes in the synthsis reaction by the lack of base pairing between the mismatched nucleotides. If a mismatch is detected, a 32 to 52 exonuclease activity is activated and the incorrect baseremoved. In most organisms, DNA polymerases function in a large complex called the replisome that contains multiple accssory subunits, such as the DNA clampor helicases. RNA-dependent DNA polymerases are a specialized class of polymerases that copy the seuence of an RNA strand into DNA.

They include reverse transcriptase, which is a viral enzyme involved in the infection of cells by retroviruses, and telomerase, which is required for the replication of telomeres. Telomerase is an unusual polymerase because it contains its own RNA template as part of its structure. Transription is carried out by a DNA-dependent RNA polymerase tat copies the sequence of a DNA strand into RNA. To begin transcribing a gene, the RNA polymrase binds to a sequence of DNA called promoter and separates the DNA strands.

It then copies the gene sequence into a messenger RNA transcript ntil it reaches a region of DNA called the terminator, where ithalts and detaches rom the DNA. s with human DNA-dependent DNA polymerases, RNA polymerase II, the enzyme that transcribes most of the genes in the human genome, operates as part of a large protein complex with multiple regulatory and accessory subunts.

Genetic Recombination

A DNA helix usually does not ineract with other segments of DNA, and in human cells the different chromosomes even occupy separate areas in the nucleus called "chromosome territories". This physical separation of different chromosomes isimportant for the ability of DNA to function as a stable repository for information, as one of the few times chromosomes interact is during chromosomal crossover when they recombine. Chroosomal crossover is when two DNA helices break, swap a section and then rejoin. Recombination allows chromosomes to exchange genetic information and produces new combinations of genes, which increases the efficiency of natural selection ad can be important in the rapid evolutio of new proteins. Genetic recombiation an also be involved in DNA repair, particularly in the cell's response to double-strand breaks. The most common form of chomosomal crossover is homologous recombination, where the two chromosomes involved share very similar sequences. Non-homologous recombination can be damaging to cells, as it can produce chromosomal translocations and genetic abnormalities. The recombination reaction is catalyzed by enzymes known as recombinases, such as RAD51.

The first step in recombination is a double-stranded break either caused by anendonuclease or damage to the DNA. A series of steps catalyzed in part by the recombinase then leads to joining of the two helices by at least one Holliday junction, in which a segment of a single strand in each helix is annealed to the complementary strand in the other helix. The Holliday junction is a tetrahedral junction structure that can be moved along the pair of chromosomes, swapping one strand for another. The recombination reaction is then halted by cleavage of the junction ad re-ligation ofthe released DNA.

Evolution

DNA contains the genetic information that allows all modern living things to function, grow and reproduce. However, it is unclear how long in the -billion-year history of life DNA has performed this function, as it has been proposed that the earliest forms of life may have used RNA as their genetic material. RNA may have acted as the central part of early cell metabolism as it can both transmit genetic information and carry out catalysis as part of ribozymes.

This ancient RNA world where nucleic acid would have been used for both catalysis and genetics may have influenced the evolution of the current genetic code ased on four nucleotide bases. This would occur, since the number of different bases in such an organism is a trade-off between a small number of bases increasing replication accuracy and a large number of bases increasing the catalytic efficiency of ribozymes. However, there is no direct evidence of ancient geetic systems, as recovery of DNA from most fossils i impossible. This is because DNA will survive in the environment for less than one million years and slowly degrades into short fragments n solution. Claims for older DNA have been made, most notably a report of the isolation of a viable bacterium from a salt crystal 250 million years old, but these claims are controversial.

Uses in Technology

Genetic Engineering

Genetic engineering, also called genetic modification, is the direct human manipulation of an organism's genetic material in a way thatdoes not occur under natural conditions. It inolves the use of recombinant DNA techniques, but does not include traditional animal and plant breeding or mutagenesis. Any organism that is generated using these techniques is considered to be a genetically modified organism. The first organisms genetically engineered wee bacteria in 1973 and then mice in 1974. Insulin producing bacteria were commercialized in 1982 and genetically modified food has been sold since 1994. Producing genetically modified organisms or tissues is a multi-step process. It first involves the isolaing and copying the genetic material of interest; if necessary, changes to genetic sequence may be introduced. A construct may be built containing all the genetic elements for correct expression in a vector.

This vector construct is then introduced into the host organism in a process called transformation, transfection or transduction. Successfully transformed organisms or tissues are then selectively grown, usually by growing material in conditions which require the presence of a gene in the inserted material for survival. Genetic engineering techniques have been applied tovarious industries, with some success. Medicines such as insulin and human gowth hormone are now produced in bacteria, experimental mice such as the oncomouse and the knockout mouse are being used for research puroses and insect resistant and/or herbicide tolerant crops have been commercialized. Plants that contain drugs and vaccines, animals with beneficial proteins in their milk and stress tolerant crops are currently being developed.

Forensics

Foensic scientists can use DNA in blood, semen, kin, saliva or hair found at a crime scen to identify a matching DNA of an individual, such as a perpetrator. This prcess is formally termed DNA profiling, but may also be called "genticfingerprinting". In DNA profiling, the lenths of variable ections of repetitive DNA, such as short tandem repeats and minisatellites, are compared between people.

This method is usually an extremely reliable technique for dentifying a matching DNA.However, identification can be complicated if the scen is contaminated with DA from several people.DNA profiling was developed in 1984 by British geneticist Sir Alec Jeffreys,and first used in forensic science to convict Colin Pitchfork in the 1988 Enderby murders case. People convicted of certain types of crimes may be required o provide a sample of DNA for a database. This has helped investigators solve ld cases where only a DNA sample was obained from the scene. DNA profiling can also be used to identify victims of mass casualty incidents.On the other hand, many convicted people have been released from rison on the basis of DNA techniques, which were not available when a crime had originally been committed.

Bioinformatics

Bioinformatics involves the manipulation, searching, and data mining of biological dta, and this includes DNA sequence data. The development of techniqus to store and search DNA sequences have led to widely applied advances in computer science, especially string searching algorithms, machine learning and database theory.String searching or matching algorithms, which find an occurrence of a sequence of letters inside a larger sequence of letters, were developed to search for specific sequencs of nucleotides. The DNA sequenced maybe aligned with other DNA sequences to identify homologous sequences and locate the specific mutations that make them distinct.

These techniques, especially multile sequence alignment, are used in studying phylogenetic relationships and protein function. Data set representing entie genomes' worth of DNA sequences, such as those produced by the Human Genome Project, are difficult to use without the annotations that identify the locations of genes and regulatory elements on each hromosome.

Regions of DNA sequence that have the characteristic patterns associated with protein- or RNA-coding genes can be identified by gene finding algorithms, which allow researchers to predict the presence of particular gene products and their possible functions in an organism even before they have been isolate experimentally.Entire genomes may also be compared which can shed light on the evolutionary history of particular organism and permit the examination of complx evolutionary event.

DNA Nanotechnology

DNA nanotechnoloy uses the uniquemolecular reconition properties of DNA and other nucleic acids to reate sel-assembling branched DNA omplexes wih useful properties.DNA is thus used as a structural material rather than as a carrier of biological information. This has led to the creation of two-dimensional periodic lattices (both tile-based s well as using the "DNA origami" method) as well as three-dimensional structures in the shapes of polyhedra. Nanomechanical devices and algorithmic self-assembly have also been dnstrated,and these DNA structures have been used to template the arrangement of other molecules such as gold nanoparticles and streptavidin proteins.

History and Athropology

Because DNA collects mutations over time, which are then inherited, it contains historical informatio, and, by comparing DNA sequences, geneticists can infer the evolutionary history of organisms, their phylogeny. This field of phylogenetics is a powerful tool in evoltionary biology. If DNA sequences within a species are compared, population geneticists can learn the history of particular populations.

This can be used in studies ranging from ecological genetics to anthroology; For example, DNA evidence is being used to try to identify the Ten Lot Tribes of Israel. DNA has also been used to look at modern fmily relationships, such as establishing family relationships between the descendants of Sally Hemings and Thomas Jefferson. This usage is closely reated to the use of DNA in criminal investigations. Indeed, some criminal investigations hae been solved when DNA from crime scenes has matched relatives of the guilty individual.

Protein Synthesis

During the 1950s and 1960s it bcame apparent that DNA is essential in the synthesis of proteins. Proteins are used as structural materials in the cells and function as enzymes. In addition, many specialized proteins function in cellular activities. For example, in bacteria, flagella and pili are composed of protein.

The Genetic Code

The key element of a protein molecule is how the amino acids are linked. The sequences of amino acids, determined y genetic codes in DNA, distinguish one protein from another. The genetic code consists of the sequence of nitroenous bases in the DNA. How the nitrogenous base code is translated to an amino acid sequence in a protein is the basis for protein synthesis. In order for protein synthesis to occur, several essential materials must be present.

One is a supply of the 20 amino acids which make up most proteins. Another essential element is a series of enzymes that will functon in the process. DNA and another form of nucleic acid called ribonucleic acid (RNA) are also essential. RNA carries instructios from the nuclear DNA into the cytoplasm, where protein is synthesized RNA is similar to DNA, with three exceptions. First, the carbohydrate in RNA is ribose rather than deoxyribose. Second, RNA nucleotides contain the pyrimidine uracil rather than thymine. And third, RNA is usually single-stranded.

Types of RNA

In the synthesis of protein, three types of RNA are required. The first is called ribosomal RNA (rRNA) and is used to manufacture ribosomes. Ribosomes are ultramicroscopic particles of rRNA and protein where amino acids are linked to one another during the synthesis of proteins. Ribosomes may exist along the membranes of the endoplasmic reticulum in eukaryotic cells or free in the cytoplasm of prokaryotic cells.

A second important type of RNA is transfer RNA (tRNA), which is used to carry amino acids to the ribosomes for protein synthesis. Molecules of tRNA exist free in the cytoplasm of cells. When protein synthesis is taking place, enzymes link tRNA to amino acids in a highly specific manner.

The third form of RNA is messengr RNA (mRNA), which receives thegenetic code from DNA and carries it into the cytoplasm where protein synthesis takes place. In this way, a eneticcode in the DNA can be used to synthesize a protein at a distant location at the ribosome. The synthesis of mRNA, tRNA, and rRNA is accomplished by an enzyme called RNA polymerase.

Transcription

Transcription is one of the first processes in the overall process of protein synthesis. In transcription, a strand of mRNA is synthesized using he genetic code of DNA. RNA polymerase bids to an area of a DNA molecule in the double helix (the other strand remains unused). The enzme moves along the DNA strand and selects complementary bases from available nucleotides and positions them in an mRNA molecule just as to the principe of complementary base pairing. The chain of mRNA lengthens until a stop code is received.

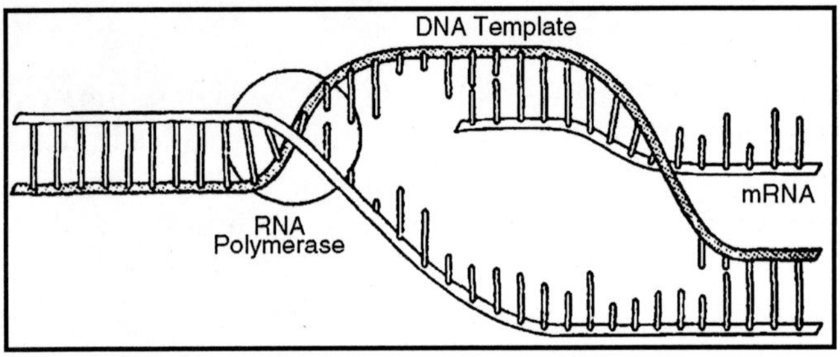

***Figure:** The Synthesis of mRN using a Strand of DNA as a Template.*

Thenucleotides of the DNA strands are read in groups of three. Each triplet is called a codon. Thus, a codon may be CGA, or TTA, or GCT, or any other combination of the four bases, depending on thir sequence in the DNA strand.

The mRNA molecule consists of a series of codons received from the gentic message in the DNA. Once the stop codon has been reached, the mRNA olecule leaves the DN molecule, and the DNA molecule rewinds to form a double helix. Meanwhile, the mRNA molecule proceeds thorough the cellular cytoplasm towards the ribosomes.

Translation

Translation is the process in which the genetic code will be "translated" to an amino acid sequence in a protein. The process begins with the arrival of the mRNA molecule at the ribosomes. While mRNA was being synthesized, tRNA molecules were uniting with their specific amino acids just as to the activity of specific enzymes. The tRNA molecules then began transporting their amino acids to the ribosomes to meet the mRNA molecule.

After it arrives at the ribosomes, the mRNA molecule exposes its bases in sets of three, the codons. Each codon has a complementary codon called an anticodon on a tRNA molecule. When the codon of the mRNA molecule complements the anticodon on a tRNA molecule, the latter places the particular amino acid in that position. Then the next codon of the mRNA is exposed, and the complementary anticodon of a tRNA molecule matches with it.

The amino acid carried by the second tRNA molecule is thus positioned next to the first amino acid, and the two are linked. At this point, the tRNA molecules release their amino acids and return to the cytoplasm to link up with new molecules of amino acid. The ribosome then moves farther down the mRNA molecule and exposes another codon which attracts another tRNA molecule with its anticodon. Another amino acid is brought into position. In this way, amino acids continue to be added to the growing chain until the ribosome has moved down to the end of the mRNA molecule. The sequence of codons on the mRNA molecule thus determines the sequence of amino acids in the protein being constructed.

Once the protein has been completely synthesized, it is emoved from the ribosome for further processing. For example, the protein may be stored in the Golgi body of a eukaryotic cell before release, or a bacterium may release it as a toxin. The mRNA molecule is broken up and the nucleotides are returned to the nucleus. The tRNA molecules return to the cytoplasm to unite with fresh molecules of amino acids, and the ribosome awaits the arrival of a new mRNA molecule.

Gene Control

The process of protein synthesis does not occur constantly in the cell, but rather at intervals followed by periods of genetic "silence." Thus, the process of gene expression is regulated and controlled by the cell. The control of gene exprssion can occur at several levels in the cell. For example, genes rarely operate during mitosis. Other levels of gene control can occur at transcription, whencertain segments of DNA increase and acelerate the activity of nearby genes.

After transcription has taken place, the mRNA molecule can be altered to regulate gene activity. For example, it has been found that eukaryotic mRNA contains many useless bits of RNA that are rmoved in the production of the final mRNA molecule. These useless bits of nucleic acid are called introns. The remaining pieces of mRNA, called exons, are then spliced to form the final mRNA molecule. Bacterial mRNA lacks introns.

The concept of gene control has been researched thoroughly in bacteria. In these microorganisms, genes have been identified as structural genes, regulator genes, and control regions. The three units form a functional unit called the operon. The operon has been examined in close detail in certain bacteria. It has been found that certain carbohydrates can induce the presence of the enzymes needed to digest those carbohydrates. For example, when lactose is present, bacteria synthesize the enzymes needed to break it down. Lactose acts as an inducer molecule in the following way: In the absence of lactose, a regulator gene produces a repressor protein, which binds to a control region called the operator site. This binding prevents the structural genes from encoding the enzyme for lactose digestion. When lactose is present, however, it binds to the repressor protein and thereby removes the repressor at the operator site. With the operator site free, the structural genes are released to produce their lactose-digeting enzyme.

Recombinant DNA and Biotechnology

Biotechnology is an industrial process that uses he scientific reseach on DNA for practical benefits. Biotechnology is synonymous with genetic engineering because the genes of an organism are changedduring the process and the DNA of the organismis recombined. Recombinant DNA and biotechnology can be used to form proteins not normally produced in a cell. In addition, bactria that carry recombinant DNA can be released into the environment to increase the fertility of the soil, serve as an insecticide, or relieve pollution.

Tools of Biotechnology

The bsic process of rcombinant DNA technology revoles around the activity of DNA in the synthesis of protein By intervening in this process, scientists ca change the nature of the DNA and of the gene make-up of an organism. By inserting genes into the genome of an organism, the scientist can induce the organism to roduce a protein it does not normally produce.Thetechnology of recombinant DNA has been made possible in part by extensive research on microorganisms during the last century. One important miroorganism in rcombinant DNA research is Escherichia coli (E. coli). The biochemistry an genetics of E. coi arewell known, and its DNA has been isolated and madto accept new genes. The DNA can then be forced into fresh cells of E. coli, and the bacteria will begi to produce the proteins specified by the foreign genes. Such altered bacteri are said to have been transformd. Interest in recombinant DNA and biotechnology heightened considerably in the 1960s and 1970s with the discovery of restriction enzymes. These enzyms catalyze the opening of a DNA molecule at a "restrictd" point, regardless of the DNA's source. Moreover, certain restriction enymes leave dangling ends of DNA moleules at the point where the DNA is open. (The most commonly used restriction enzme is named EcoRl.) Foreign DNA can then e combined with the carrier DNA at his point. An enzyme called DNA ligase is used to form a permanent link betwen the dangling ends of the DNA molecules at the point of union.

Pharmaceutical Products

Gene defects in humans can lead to deficiencies in proteins such as insulin, human growth hormone, and Factor VIII. These protein deficiencies may lead to problems such as diabetes, dwarfism, and impaired blood clotting, respectively. Missing proteins can now be replaced by proteins manufactured through biotechnology. For insulin production, two protein chains are encoded by separate genes in plasmids inserted into bacteria. The protein chains are then chemically joined to form the final insulin product. Human growth hormone is also produced within bacteria, but special techniques are used because the bacteria do not usually produce human proteins. Therapeutic proteins produced by biotechnology include a clot-dissolving protein called tissue plasminogen activator (TPA) and interferon.

This antiviral protein is produced within *E. coli* cells. Interferon is currently used against certain types of cancers and for certain skin conditions. Vaccines represent anothr application of recombinant DNA technology. For instance, the hepatitis B vaccine nowin use is composed of viral protein manufactured by yeast cells, which have been recombined with viral genes. The vaccine is safe because it contains no viral particles. Experimental vaccines against AIDS are being produced in the same way.

Dagnostic Testing

Recombinant DNA and biotechnology have opened a new era of diagnostic testing and have made detecting many genetic diseass possible. The basic tool of DNA analyses isa fragment ofDNA called the DNA probe. A DNA probe is a relatively smal, single-stranded fragment of DNA hat recognizes and bindsto a complementary section o DNA in a complex mixture of DNA molecules. The proe mingles with the mixture of NA and unites with the target DNA much like a left hand unites with the right. Once the probe unites with its target, it emits a signal such as radioactivity to indicate that a reaction has occurred. To work effectively, a suffiiently large amount of target DNA must be available. To inrease the amount of available DNA, a process called the polymerase chain reaction (PCR) is used.

In a highlyautomated machine, the target DNA is combined with enzyes, nucleotides, and a primer DNA. In geometric fashion, the enzymes snthesize copies of the target DNA, so that in a few ours billions of molecules of DNA exist whereonly a few were before. Using DNA probes and PCR, scientits are now able to detect the DNA associated with HIV (and AIDS), Lyme disease, and genetic diseases such as cystic fibrosis, muscular dystrophy, Huntington's disease, and fragile X syndrome.

Gene Therapy

Gene therapy is a recombinant DNA process in which cells are taken from the patient, altered by adding genes, and replaced in the patient, where the genes provide the genetic codes for proteins the patient is lacking. In the early 1990s, gene therapy was used to correct a deficiency of the enzyme adenosine deaminase (ADA).

Blood cells called lymphocytes were removed from the bone marrow of two children; then genes for ADA production were inserted into the cells using viruses as vectors. Finally, the cells were reinfused to the bodies of the two children. Once established in the bodies, the gene-altered cells began synthesizing the enzyme ADA and alleviated the deficiency. Gene therapy has also been performed with patients with melanoma (a virulent skin cancer). In this case, lymphocytes that normally attack tumors are isolated in the patients and treated with genes for an anticancer protein called tumor necrosis factor. The genealtered lymphocytes are then reinfused to the patients, where they produce the new protein which helps destroy cancer cells. Approximately 2000 single-gene defects are believed to exist, and patients with these defects my be candidates for gene therap.

DNA Fingerprinting

The use of DNA probes and the development of retrieval techniqes have made it possible to match DNA molecules to one another for identification purposes. This process has been ued in a forensic procedure calld DNA fingerprinting. The use of DNA fingerprinting depends upon the presence of repeating base sequences that exist in the human genome.

The repeating sequences are called restriction fragment length polymorphisms (RFLPs). As the pattern of RFLPs is unique for every individual, it can be used as amolecular fingerprin. To perform DNA fingerprinting, DNA is obtained from an individual's blood cells, hair fibres, skn fragments, or other tissue. The DNA is extracted from the cells and digested with enzymes. The resulting fragments are separated by a process calle electrophoresis. These separated DNA fragments are tesed for characteristic RFLPs using DNA probes. A statistical evaluation enables the forensic athologist tocompare a suspect's DNA with the DNA recovered at a crime scene and to assert with a degree of certainty (usually 99 per cent) that te suspect was at the crime scene.

DNA and Agriculture

Although plants are more difficult to work with than bacteria, gene insertions can be made into single plant cells, and the cells can then be cultivated to form a mature plant. The major method for inserting genes is through the plasmids of a bacterium called Agrobacterium tumefaciens. This bacterium invades plant cells, and its plasmids insert into plant chromosomes carrying the genes for tuor induction.

Scientists remove the tumor-inducing genes and obtain a plasmid that unites with the plant cell wthout causing any harm. Recombinant DNA and biotechnology have been used to increase the efficiency of plant growth by increasing the efficiency of the plant's ability to fix nitrogen. Scientists have obtained the genes for nitrogen fixation from bacteria and have incorporated tose genes into plant cells.

By obtaining nitrogen directly from the atmosphere, the plants can synthesize their own proteins without intervntion of bacteria as normally needed. DNA technology has also been used to increase plant resistance to disease. The genes for an insecticide have been obtained from the bacterium Bacillus thuringiensis and inserted into plants to allow them to resist caterpillars and other pests. In addition, plants have been reengineered to produce the capsid protein that encloses viruses. These proteins lend resistance to the plants against viral disease.

The Human Genome

One of the most ambitious scientific endeavors of the twentieth century was the effort to sequence the nitrogenous bases in the human genome. Begun in 1990 and completed in 2003, the effort encompassed 13 years of work at a cost of approximately $3 billion. Knowing the content of the human genome is helping researchers devise new diagnostics and treatments for genetic diseases and will also be of value to developmental biologists, evolutionary biologists, and comparative biologists. In addition to learning the genome of humans, the project has also tudied numerous bacteria. By 1995, the genomes of two bacteria had been completely deciphered (Haemophilus influenzae and Mycoplasma genitalium), and by 1996, the genome of the yeast Saccharomyces cerevisiae was known.

The Human Genome Project is one of colossal magnitude that will have an impact on many branches of science for decades to come. The projct remains the crowning achievement of DNA research in the twentieth century and the bedrock for research in the twenty-first.

4

Genetic Material: Properties and Replication

DNA

Deoxyribonucleic acid (DNA) is a nucleic acid that contains the genetic instructions used in the development and functioning of all known living organisms with the exception of some viruses. The main role of DNA molecules is the long-term storage of information. DNA is often compared to a set of blueprints, like a recipe or a code, since it contains the instructions needed to construct other components of cells, such as proteins and RNA molecules. The DNA segments that carry this genetic information are called genes, but other DNA sequences have structural purposes, or are involved in regulating the use of this genetic information.

DNA consists of two long polymers of simple units called nucleotides, with backbones made of sugars and phosphate groups joined by ester bonds. These two strands run in opposite directions to each other and are therefore anti-parallel. Attached to each sugar is one of four types of molecules called bases. It is the sequence of these four bases along the backbone that encodes information.

This information is read using the genetic code, which specifies the sequence of the amino acids within proteins. The code is read by copying stretches of DNA into the related nucleic acid RNA, in a process called transcription. Within cells, DNA is organized into long structures called chromosomes. These chromosomes are duplicated before cells divide, in a process called DNA replication. Eukaryotic organisms (animals, plants, fungi, and protists) store most of their DNA inside the cell nucleus and some of their DNA in organelles, such as mitochondria or chloroplasts. In contrast, prokaryotes (bacteria and archaea store their DNA ony in the cytoplasm. Within the chromosomes, chromatin proteins such as histones compact and organize DNA. These compact structures guide the interactions between DNA and other proteins, helping control which parts of the DNA are transcribed.

Properties

DNA is a long polymer made from repeating units called nucleotides. The DNA chain is 22 to 26 Angströms wide (2. to 2.6 nanometres), and one nucleotide unit is 3.3 Å (0.33 nm) long. Although each individual repeating unit is very small, DNA polymers can be very large molecules containing millions of nucleotides. For instance, the largest human chromosome, chromosome number 1, is approximately 220 million base pairs long. In living organisms, DNA does not usually exist as a single molecule, but instead as a pair of molecules that are held tightly together. These two long strands entwine like vines, in the shape of a double helix.

The nucleotide repeats contain both the sgment of thebackbone of the molecule, which holds the chain together, and a base, which interacts with the other DNA strand in the helix. A base linked to a sugar is called a nucleoside and a base linked to a sugar and one or more phosphate groups is called a nucleotide. If multiple nucleotides are linked together, as in DNA, this polymer is called a polynucleotide.

The backbone of the DNA strand is made from alternating phosphate and sugar residues. The sugar in DNA is 2-deoxyribose, which is a pentose (five-carbon) sugar. The sugars are joined together by phosphate groups that form phosphodiester bonds between the third and fifth carbon atoms of adjacent sugar rings. These asymmetric bonds mean a strand of DNA has a direction. In a double helix the direction of the nucleotides in one strand is opposite to their direction in the other strand: the strands are *antiparallel*. The asymmetric ends of DNA strands are called the 52 (*five prime*) and 32 (*three prime*) ends, with the 5' end having a terminal phosphate group and the 3' end a terminal hydroxyl group. One major difference between DNA and RNA is the sugar, with the 2-deoxyribose in DNA being replaced by the alternative pentose sugar ribose in RNA.

The DNA double helix is stabilized by hydrogen bonds between the bases attached to the two strands. The four bases found in DNA are adenine (abbreviated A), cytosine (C), guanine (G) and thymine (T). These four bases are attached to the sugar/phosphate to form the complete nucleotide, as shown for adenosine monophosphate.

These bases are classified into two types; adenine and guanine are fused five- and six-membered heterocyclic compounds called purines, while cytosine and thymine are six-membered rings called pyrimidines. A fifth pyrimidine base, called uracil (U), usually takes the place of thymine in RNA and differs from thymine by lacking a methyl group on its ring. Uracil is not usually found in DNA, occurring only as a breakdown product of cytosine. In addition to RNA and DNA, a large number of artificial nucleic acid analogues have also been created to study the proprieties of nucleic acids, or for use in biotechnology.

Grooves

Twin helical strands form the DNA backbone. Another double helix may be found by tracing the spaces, or grooves, between the strands. These voids are adjacent to the base pairs and may provide a binding site. As the strands are not directly opposite each other, the grooves are unequally sized. One groove, the major groove, is 22 wide and the other, the minor groove, is 12 wide. The narrowness of the minor groove means that the edges of the bases are more accessible in the major groove. As a result, proteins like transcription factors that can bind to specific sequences in double-stranded DNA usually make contacts to the sides of the bases exposed in the major groove.

This situation varies in unusual conformations of DNA within the cell, but the major and minor grooves are always named to reflect the differences in size that would be seen if the DNA is twisted back into the ordinary B form.

Base Pairing

Each type of base on one strand forms a bond with just one type of base on the other strand. This is called complementary base pairing. Here, purines form hydrogen bonds to pyrimidines, with A bonding only to T, and C bonding only to G. This arrangement of two nucleotides binding together across the double helix is called a base pair.

As hydrogen bonds are not covalent, they can be broken and rejoined relatively easily. The two strands of DNA in a double helix can therefore be pulled apart like a zipper, either by a mechanical force or high temperature. As a result of this complementarity, all the information in the double-stranded sequence of a DNA helix is duplicated on each strand, which is vital in DNA replication. Indeed, this reversible and specific interaction between complementary base pairs is critical for all the functions of DNA in living organisms.

The two types of base pairs form different numbers of hydrogen bonds, AT forming two hydrogen bonds, and GC forming three hydrogen bonds. DNA with high GC-content is more stable than DNA with low GC-content, but contrary to popular belief, this is not due to the extra hydrogen bond of a GC base pair but rather the contribution of stacking interactions (hydrogen bonding merely provides specificity of the pairing, not stability). As a result, it is both the percentage of GC base pairs and the overall length of a DNA double helix that determine the strength of the association between the two strands of DNA. Long DNA helices with a high GC content have stronger-interacting strands, while short helices with high AT content have weaker-interacting strands. In biology, parts of the DNA double helix that need to separate easily, such as the TATAAT Pribnow box in some promoters, tend to have a high AT content, making the strands easier to pull apart. In the laboratory, the strength of this interaction can be measured by finding the temperature required to break the hydrogen bonds, their melting temperature (also called T_m value). When all the base pairs in a DNA double helix melt, the strands separate and exist in solution as two entirely independent molecules. These single-stranded DNA molecules (*ssDNA*) have no single common shape, but some conformations are more stable than others.

Sense and Antisense

A DNA sequence is called "sense" if its sequence is the same as that of a messenger RNA copy that is translated into protein. The sequence on the opposite strand is called the "antisense" sequence. Both sense and antisense sequences can exist on different parts of the same strand of DNA (i.e. both strands contain both sense and antisense sequences). In both prokaryotes and eukaryotes, antisense RNA sequences are produced, but the functions of these RNAs are not entirely clear. One proposal is that antisense RNAs are involved in regulating gene expression through RNA-RNA base pairing.

A few DNA sequences in prokaryotes and eukaryotes, and more in plasmids and viruses, blur the distinction between sense and antisense strands by having overlapping genes. In these cases, some DNA sequences do double duty, encoding one protein when read along one strand, and a second protein when read in the opposite direction along the other strand. In bacteria, this overlap may be involved in the regulation of gene transcription, while in viruses, overlapping genes increase the amount of information that can be encoded within the small viral genome.

Supercoiling

DNA can be twisted like a rope in a process called DNA supercoiling. With DNA in its "relaxed" state, a strand usually circles the axis of the double helix once every 10.4 base pairs, but if the DNA is twisted the strands become more tightly or more loosely wound. If the DNA is twisted in the direction of the helix, this is positive supercoiling, and the bases are held more tightly together. If they are twisted in the opposite direction, this is negative supercoiling, and the bases come apart more easily. In nature, most DNA has slight negative supercoiling that is introduced by enzymes called topoisomerases. These enzymes are also needed to relieve the twisting stresses introduced into DNA strands during processes such as transcription and DNA replication.

Alternate DNA Structures

DNA exists in many possible conformations that include A-DNA, B-DNA, and Z-DNA forms, although, only B-DNA and Z-DNA have been directly observed in functional organisms. The conformation that DNA adopts depends on the hydration level, DNA sequence, the amount and direction of supercoiling, chemical modifications of the bases, the type and concentration of metal ions, as well as the presence of polyamines in solution.

The first published reports of A-DNA X-ray diffraction patterns— and also B-DNA used analyses based on Patterson transforms that provided only a limited amount of structural information for oriented fibres of DNA. An alternate analysis was then proposed by Wilkins *et al.*, in 1953, for the *in vivo* B-DNA X-ray diffraction/scattering patterns of highly hydrated DNA fibres in terms of squares of Bessel functions. In the same journal, James D. Watson and Francis Crick presented their molecular modeling analysis of the DNA X-ray diffraction patterns to suggest that the structure was a double-helix.

Although the 'B-DNA form' is most common under the conditions found in cells, it is not a well-defined conformation but a family of related DNA conformations that occur at the high hydration levels present in living cells. Their corresponding X-ray diffraction and scattering patterns are characteristic of molecular paracrystals with a significant degree of disorder.

Compared to B-DNA, the A-DNA form is a wider right-handed spiral, with a shallow, wide minor groove and a narrower, deeper major groove. The A form occurs under non-physiological conditions in partially dehydrated samples of DNA, while in the cell it may be produced in hybrid pairings of DNA and RNA strands, as well as in enzyme-DNA complexes. Segments of DNA where the bases have been chemically modified by methylation may undergo a larger change in conformation and adopt the Z form. Here, the strands turn about the helical axis in a left-handed spiral, the opposite of the more common B form. These unusual structures can be recognized by specific Z-DNA binding proteins and may be involved in the regulation of transcription.

Quadruplex Structures

At the ends of the linear chromosomes are specialized regions of DNA called telomeres. The main function of these regions is to allow the cell to replicate chromosome ends using the enzyme telomerase, as the enzymes that normally replicate DNA cannot copy the extreme 32 ends of chromosomes. These specialized chromosome caps also help protect the DNA ends,and stop the DNA repair systems in the cell from treating them as damage to be corrected. In human cells, telomeres are usually legths of single-stranded DNA containing several thousand repeats of a simple TTAGGG sequence.

These guanine-rich sequences my stabilize chromosome ends byforming structures of stacked sets of four-base units, rather than the usual base pairs found in other DNA molecules. Here, four guanine bases form a flat plate and these flat four-base units then stack on top of each other, to form a stable *G-quadruplex* structure.

These structures are stabilized by hydrogen bondig between the edges of the bases and chelation of a metal ion in the centre of each four-base unit. Other structures can also be formed, with the central set of four bases coming from either a single strand folded around the bases, or several different parallel strands, each contributing one base to the central structure.

In addition to these stacked structures, telomeres also form large loop structures called telomere loops, or T-loops. Here, the single-stranded DNA curls around in a long circle stabilized by telomere-binding proteins. At the very end of the T-loop, the single-stranded telomere DNA is held onto a region of double-stranded DNA by the telomere strand disrupting the double-helical DNA and base pairing to one of the two strands. This triple-stranded structure is called a displacement loop or D-loop.

Branched DNA

In DNA fraying occurs when non-complementary regions exist at the end of an otherwise complementary double-strand of DNA. However, branched DNA can occur if a third strand of DNA is introduced and contains adjoining regions able to hybridize with the frayed regions of the pre-existing double-strand. Although the simplest example of branched DNA involves only three strands of DNA, complexes involving additional strands and multiple branches are also possible.

Vibration

DNA may carry out low-frequency collective motion as observed by the Raman spectroscopy and analyzed with a quasi-continuum model.

DNA-binding Proteins

Structural proteins that bind DNA are well-understood examples of non-specific DNA-protein interactions. Within chromosomes, DNA is held in complexes with structural proteins. These proteins organize the DNA into a compact structure called chromatin. In eukaryotes this structure involves DNA binding to a complex of small basic proteins called histones, while in prokaryotes multiple types of proteins are involved. The histones form a diskshapd complex called a nucleosome, which contains two complete turns of double-stranded DNA wrapped around its surface. These non-specific interactions are formed through basic residues in the histones making ionic bonds to the acidic sugar-phosphate backbone of the DNA, and are therefore largely independent of the base sequence. Chemical modifications of these basic amino acid residues include methylation, phosphorylation and acetylation.

These chemical changes alter the strength of the interaction between the DNA and the histones, making the DNA more or less accessible to transcription factors and changing the rate of transcription. Other non-specific DNA-binding proteins in chromatin include the high-mobility group proteins, which bind to bent or distorted DNA. These proteins are important in bending arrays of nucleosomes and arranging them into the larger structures that make up chromosomes.

A distinct group of DNA-binding proteins are the DNA-binding proteins that specifically bind single-stranded DNA. In humans, replication protein A is the best-understood member of this family and is used in processes where the double helix is separated, including DNA replication, recombination and DNA repair.

These binding proteins seem to stabilize single-stranded DNA and protect it from forming stem-loops or being degraded by nucleases. In contrast, other proteins have evolved to bind to particular DNA sequences. The most intensively studied of these are the various transcription factors, which are proteins that regulate transcription. Each transcription factor binds to one particular set of DNA sequences and activates or inhibits the transcription of genes that have these sequences close to their promoters.

The transcription factors do this in two ways. Firstly, they can bind the RNA polymerase responsible for transcription, either directly or through other mediator proteins; this locates the polymerase at the promoter and allows it to begin transcription. Alternatively, transcription factors can bind enzymes that modify the histones at the promoter; this will change the accessibility of the DNA template to the polymerase.

As these DNA targets can occur throughout an organism's genome, changes in the activity of one type of transcription factor can affect thousands of genes. Consequently, these proteins are often the targets of the signal transduction processes that control responses to environmental changes or cellular differentiation and development.

The specificity of these transcription factors' interactions with DNA come from the proteins making multiple contacts to the edges of the DNA bases, allowing them to "read" the DNA sequence. Most of these base-interactions are made in the major groove, where the bases are most accessible

Uses in Technology

Genetic Engineering

Methods have ben developed to purify DNA from organisms, such as phenol-chloroform extraction and manipulate it in the laboratory, such as estriction digests and the polymerase chain reaction. Modern biology and biochemistry make intensive use of these techniques in recombinant DNA technology. Recombinant DNA is a manmade DNA sequence that has been assembled from other DNA sequences. They can be transformed into organisms in the form of plasmids or in the appropriate format, by using a viral vector. The genetically modified organisms produced can be used to produce products such as recombinant proteins, used in medical research, or be grown in agriculture.

Forensics

Forensic scientists can use DNA in blood, semen, kin, saliva or hair found at a crime scene to identify a matching DNA of an individual, such as a perpetrator. This process is formally termed DNA profiling, but may also be called "genetic fingerprinting". In DNA profiling, the lengths of variable sections of repetitive DNA, such as short tandem repeats and minisatellites, are compared between people.

This method is usually an extremely reliable technique for identifying a matching DNA. However, identification can be complicated if the scene is contaminated with DNA from several people. DNA profiling was developed in 1984 by British geneticist Sir Alec Jeffreys, and first used in forensic science to convict Colin Pitchfork in the 1988 Enderby murders case.

Pople convicted of certain types of crimes may be required to provide a sample of DNA for a database. This has helped investigators solve old cases where only a DNA sample was obtained from the scene. DNA profiling can also be used to identify victims of mass casualty incidents. On the other hand, many convicted people have been released from prison on the basis of DNA techniques, which were not available when a crime had originally been committed.

DNA Profiling

DNA profiling (also called DNA testing, DNA typing, or genetic fingerprinting) is a technique employed by forensic scientists to assist in the identification of individuals on the basis of their respective DNA profiles. DNA profiles are encrypted sets of numbers that reflect a person's DNA make-up, which can also be used as the person's identifier. DNA profiling should not be confused with full genome sequencing. It is used in, for example, parental testing and rape investigation.

Although 99.9% of human DNA sequences are the same in every person, enough of the DNA is different to distinguish one individual from another. DNA profiling uses repetitive ("repeat") sequences that are highly variable, called variable number tandem repeats (VNTR). VNTRs loci are very similar between closely related humans, but so variable that unrelated individuals are extremely unlikely to have the same VNTRs.

The DNA profiling technique was first reported in 1984 by Sir Alec Jeffreys at the University of Leicester in England, and is now the basis of several national DNA databases. Dr. Jeffreys's genetic fingerprinting was made commercially available in 1987, when a chemical company, ICI, started a blood-testing centre in England.

DNA Profiling Process

The process begins with a sample of an individual's DNA (typically called a "reference sample"). The most desirable method of collecting a reference sample is the use of a buccal swab, as this reduces the possibility of contamination. When this is not available (e.g. because a court order may be needed and not obtainable) other methods may need to be used to collect a sample of blood, saliva, semen, or other appropriate fluid or tissue from personal items (e.g. toothbrush, razor, etc.) or from stored samples (e.g. banked sperm or biopsy tissue). Samples obtained from blood relatives (biological relative) can provide an indication of an individual's profile, as could human remains which had been previously profiled. A reference sample is then analyzed to create the individual's DNA profile using one of a number of techniques, discussed below. The DNA profile is then compared against another sample to determine whether there is a genetic match.

RFLP Analysis

The first methods for finding out genetics used for DNA profiling involved restriction enzyme digestion, followed by Southern blot analysis. Although polymorphisms can exist in the restriction enzyme cleavage sites, more commonly the enzymes and DNA probes were used to analyze VNTR loci. However, the Southern blot technique is laborious, and requires large amounts of undegraded sample DNA. Also, Karl Brown's original technique looked at many minisatellite loci at the same time, increasing the observed variability, but making it hard to discern individual alleles (and thereby precluding parental testing). These early techniques have been supplanted by PCR-based assays.

PCR Analysis

With the invention of the polymerase chain reaction (PCR) technique, DNA profiling took huge strides forward in both discriminating power and the ability to recover information from very small (or degraded) starting samples. PCR greatly amplifies the amounts of a specific region of DNA, using oligonucleotide primers and a thermostable DNA polymerase.

Early assays such as the HLA-DQ alpha reverse dot blot strips grew to be very popular due to their ease of use, and the speed with which a result could be obtained. However they were not as discriminating as RFLP. It was also difficult to determine a DNA profile for mixed samples, such as a vaginal swab from a sexual assault victim.

Fortunately, the PCR method is readily adaptable for analyzing VNTR loci. In the United States the FBI has standardized a set of 13 VNTR assays for DNA typing, and has organized the CODIS database for forensic identification in criminal cases. Similar assays and databases have been set up in other countries. Also, commercial kits are available that analyze single-nucleotide polymorphisms (SNPs). These kits use PCR to amplify specific regions with known variations and hybridize them to probes anchored on cards, which results in a coloured spot corresponding to the particular sequence variation.

STR Analysis

The method of DNA profiling used today is based on PCR and uses short tandem repeats (STR). This method uses highly polymorphic regions that have short repeated sequences of DNA (the most common is 4 bases repeated, but there are other lengths in use, including 3 and 5 bases). Because unrelated people almost certainly have different numbers of repeat units, STRs can be used to discriminate between unrelated individuals. These STR loci (locations on a chromosome) are targeted with sequence-specific primers and amplified using PCR. The DNA fragments that result are then separated and detected using electrophoresis. There are two common methods of separation and detection, capillary electrophoresis (CE) and gel electrophoresis.

Each STR is polymorphic, however, the number of alleles is small. Typically each STR allele will be shared by around 5 - 20% of individuals. The power of STR analysis comes from looking at multiple STR loci simultaneously. The pattern of alleles can identify an individual quite accurately. Thus STR analysis provides an excellent identification tool. The more STR regions that are tested in an individual the more discriminating the test becomes.

From country to country, different STR-based DNA-profiling systems are in use. In North America, systems which amplify the CODIS 13 core loci are almost universal, while in the UK the SGM+ system (which is compatible with The National DNA Database), is in use. Whichever system is used, many of the STR regions used are the same. These DNA-profiling systems are based on multiplex reactions, whereby many STR regions will be tested at the same time.

The true power of STR analysis is in its statistical power of discrimination. Because the 13 loci that are currently used for discrimination in CODIS are independently assorted (having a certain number of repeats at one locus doesn't change the likelihood of having any number of repeats at any other locus), the product rule for probabilities can be applied. This means that if someone has the DNA type of ABC, where the three loci were independent, we can say that the probability of having that DNA type is the probability of having type A times the probability of having type B times the probability of having type C. This has resulted in the ability to generate match probabilities of 1 in a quintillion (1 with 18 zeros after it) or more.

However, DNA database searches showed much more frequent than expected false DNA matches including one perfect 13 locus match out of only 30,000 DNA samples in Maryland in January 2007.

Moreover, since there are about 12 million monozygotic twins on Earth, that theoretical probability is useless. For example, the actual probability that 2 random people have the same DNA depends on whether there were twins or triplets (etc.) in the family, and the number of loci used in the test. Where twins are common, the probability of matching the DNA is 22 in 1000, or about 2.2 in 100 will have matching DNA.

In practice, the risk of contaminated-matching is much greater than matching a distant relative, such as a sample being contaminated from nearby objects, or from left-over cells transferred from a prior test. Logically, the risk is greater for matching the most common person in the samples: everything collected from, or in contact with, a victim is a major source of contamination for any other samples brought into a lab. For that reaso, multiple control-samples are typically tested, to ensure that they stayed clean, when prepared during the same period as the actual test samples. Unexpected matches (or variations) in several control-samples indicates a high probability of contamination for the actual test samples. In a relationship test, the full DNA profiles should differ (except for twins), to prove that a person wasn't actually matched as being related to their own DNA in another sample.

AmpFLP

Another technique, AmpFLP, or amplified fragment length polymorphism was also put into practice during the early 1990s. This technique was also faster than RFLP analysis and used PCR to amplify DNA samples. It relied on variable number tandem repeat (VNTR) polymorphisms to distinguish various alleles, which were separated on a polyacrylamide gel using an allelic ladder (as opposed to a molecular weight ladder). Bands could be visualized by silver staining the gel. One popular locus for fingerprinting was the D1S80 locus.

As with all PCR based methods, highly degraded DNA or very small amounts of DNA may cause allelic dropout (causing a mistake in thinking a heterozygote is a homozygote) or other stochastic effects. In addition, because the analysis is done on a gel, very high number repeats may bunch together at the top of the gel, making it difficult to resolve. AmpFLP analysis can be highly automated, and allows for easy creation of phylogenetic trees based on comparing individual samples of DNA. Due to its relatively low cost and ease of set-up and operation, AmpFLP remains popular in lower income countries.

DNA Family Relationship Analysis

Using PCR technology, DNA analysis is widely applied to determine genetic family relationships such as paternity, maternity, siblingship and other kinships.

During conception, the father's sperm cell and the mother's egg cell, each containing half the amount of DNA found in other body cells, meet and fuse to form a fertilized egg, called a zygote. The zygote contains a complete set of DNA molecules, a unique combination of DNA from both parents. This zygote divides and multiplies into an embryo and later, a full human being.

At each stage of development, all the cells forming the body contain the same DNA—half from the father and half from the mother. This fact allows the relationship testing to use all types of all samples including loose cells from the cheeks collected using buccal swabs, blood or other types of samples.

While a lot of DNA contains information for a certain function, there is some called junk DNA, which is currently used for human identification. At some special locations (called loci) in the junk DNA, predictable inheritance patterns were found to be useful in determining biological relationships. These locations contain specific DNA markers that DNA scientists use to identify individuals. In a routine DNA paternity test, the markers used are Short Tandem Repeats (STRs), short pieces of DNA that occur in highly differential repeat patterns among individuals.

Each person's DNA contains two copies of these markers—one copy inherited from the father and one from the mother. Within a population, the markers at each person's DNA location could differ in length and sometimes sequence, depending on the markers inherited from the parents.

The combination of marker sizes found in each person makes up his/her unique genetic profile. When determining the relationship between two individuals, their genetic profiles are compared to see if they share the same inheritance patterns at a statistically conclusive rate. For example, the following sample report from this commercial DNA paternity testing laboratory Universal Genetics signifies how relatedness between parents and child is identified on those special markers:

DNA Marker	*Mother*	*Child*	*Alleged father*
D21S11	28, 30	28, 31	29, 31
D7S820	9, 10	10, 11	11, 12
TH01	14, 15	14, 16	15, 16
D13S317	7, 8	7, 9	8, 9
D19S433	14, 16.2	14, 15	15, 17

The partial results indicate that the child and the alleged father's DNA match among these five markers. The complete test results show this correlation on 16 markers between the child and the tested man to draw a conclusion of whether or not the man is the biological father. Scientifically, each marker is assigned with a Paternity Index (PI), which is a statistical measure of how powerfully a match at a particular marker indicates paternity.

The PI of each marker is multiplied with each other to generate the Combined Paternity Index (CPI), which indicates the overall probability of an individual being the biological father of the tested child relative to any random man from the entire population of the same race. The CPI is then converted into a Probability of Paternity showing the degree of relatedness between the alleged father and child.

The DNA test report in other family relationship tests, such as grandparentage and siblingship tests, is similar to a paternity test report. Instead of the Combined Paternity Index, a different value, such as a Siblingship Index, is reported.

The report shows the genetic profiles of each tested person. If there are markers shared among the tested individuals, the probability of biological relationship is calculated to determine how likely the tested individuals share the same markers due to a blood relationship.

Y-chromosome Analysis

Recent innovations have included the creation of primers targeting polymorphic regions on the Y-chromosome (Y-STR), which allows resolution of a mixed DNA sample from a male and female and/or cases in which a differential extraction is not possible. Y-chromosomes are paternally inherited, so Y-STR analysis can help in the identification of paternally related males. Y-STR analysis was performed in the Sally Hemings controversy to determine if Thomas Jefferson had sired a son with one of his slaves.

Mitochondrial Analysis

For highly degraded samples, it is sometimes impossible to get a complete profile of the 13 CODIS STRs. In these situations, mitochondrial DNA (mtDNA) is sometimes typed due to there being many copies of mtDNA in a cell, while there may only be 1-2 copies of the nuclear DNA. Forensic scientists amplify the HV1 and HV2 regions of the mtDNA, then sequence each region and compare single-nucleotide differences to a reference. Because mtDNA is maternally inherited, directly linked maternal relatives can be used as match references, such as one's maternal grandmother's daughter's son.

A difference of two or more nucleotides is generally considered to be an exclusion.

Heteroplasmy and poly-C differences may throw off straight sequence comparisons, so some expertise on the part of the analyst is required. mtDNA is useful in determining clear identities, such as those of missing people when a maternally linked relative can be found. mtDNA testing was used in determining that Anna Anderson was not the Russian princess she had claimed to be, Anastasia Romanov. mtDNA can be obtained from such material as hair shafts and old bones/teeth.

DNA Databases

There are now several DNA databases in existence around the world. Some are private, but most of the largest databases are government controlled. The United States maintains the largest DNA database, with the Combined DNA Index System, holding over 5 million records as of 2007.

The United Kingdom maintains the National DNA Database (NDNAD), which is of similar size, despite the UK's smaller population. The size of this database, and its rate of growth, is giving concern to civil liberties groups in the UK, where police have wide-ranging powers to take samples and retain them even in the event of acquittal.

The U.S. Patriot Act of the United States provides a means for the U.S. government to get DNA samples from other countries if they are either a division of, or head office of, a company operating in the U.S. Under the act, the American offices of the company can't divulge to their subsidiaries/offices in other countries the reasons that these DNA samples are sought or by whom.

When a match is made from a National DNA Databank to link a crime scene to an offender who has provided a DNA Sample to a databank that link is often referred to as a *cold hit*. A cold hit is of value in referring the police agency to a specific suspect but is of ess evidential value than a DNA match made from outside the DNA Databank.

Considerations when Evaluating DNA Evidenc

In the early days of the use of genetic fingerprinting as criminal evidence, juries were often swayed by spurious statistical arguments by defence lawyers long these lines: given a match that had a 1 in 5 million probability of occurring by chance, the lawyer would argue that this meant that in a country of say 60 million people there were 12 people who would also match the profile. This was then translated to a 1 in 12 chance of the suspect being the guilty one.

This argument is not sound unless the suspect was drawn at random from the population of the country. In fact, a jury should consider how likely it is that an individual matching the genetic profile would also have been a suspect in the case for other reasons. Another spurious statistical argument is based on the false assumption that a 1 in 5 million probability of a match automatically translates into a 1 in 5 million probability of innocence and is known as the prosecutor's fallacy.

When using RFLP, the theoretical risk of a coincidental match is 1 in 100 billion (100,000,000,000), although the practical risk is actually 1 in 1000 because monozygotic twins are 0.2% of the human population. Moreover, the rate of laboratory error is almost certainly higher than this, and often actual laboratory procedures do not reflect the theory under which the coincidence probabilities were computed.

For example, the coincidence probabilities may be calculated based on the probabilities that markers in two samples have bands in *precisely* the same location, but a laboratory worker may conclude that similar—but not precisely identical—band patterns result from identical genetic samples with some imperfection in the agarose gel. However, in this case, the laboratory worker increases the coincidence risk by expanding the criteria for declaring a match.

Recent studies have quoted relatively high error rates which may be cause for concern. In the early days of genetic fingerprinting, the necessary population data to accurately compute a match probability was sometimes unavailable. Between 1992 and 1996, arbitrary low ceilings were controversially put on match probabilities used in RFLP analysis rather than the higher theoretically computed ones. Today, RFLP has become widely disused due to the advent of more discriminating, sensitive and easier technologies.

STRs do not suffer from such subjectivity and provide similar power of discrimination (1 in 10^13 for unrelated individuals if using a full SGM+ profile) It should be noted that figures of this magnitude are not considered to be statistically supportable by scientists in the UK, for unrelated individuals with full matching DNA profiles a match probability of 1 in a billion is considered statistically supportable (Since 1998 the DNA profiling system supported by The National DNA Database in the UK is the SGM+ DNA profiling system which includes 10 STR regions and a sex indicating test.

However, with any DNA technique, the cautious juror should not convict on genetic fingerprint evidence alone if other factors raise doubt. Contamination with other evidence (secondary transfer) is a key source of incorrect DNA profiles and raising doubts as to whether a sample has been adulterated is a favorite defence technique. More rarely, chimerism is one such instance where the lack of a genetic match may unfairly exclude a suspect.

Evidence of Genetic Relationship

It's also possible to use DNA profiling as evidence of genetic relationship, but testing that shows no relationship isn't absolutely certain. While almost all individuals have a single and distinct set of genes, rare individuals, known as "chimeras", have at least two different sets of genes. There have been several cases of DNA profiling that falsely "proved" that a mother was unrelated to her children.

Fake DNA Evidence

The value of DNA evidence has to be seen in light of recent cases where criminals planted fake DNA samples at crime scenes. In one case, a criminal even planted fake DNA evidence in his own body: Dr. John Schneeberger raped one of his sedated patients in 1992 and left semen on her underwear.

Police drew what they believed to be Schneeberger's blood and compared its DNA against the crime scene semen DNA on three occasions, never showing a match. It turned out that he had surgically inserted a Penrose drain into his arm and filled it with foreign blood and anticoagulants.

In a study conducted by the life science company Nucleix and published in the journal Forensic Science International, scientists found that an In vitro synthesized sample of DNA matching any desired genetic profile can be constructed using standard molecular biology techniques without obtaining any actual tissue from that person.

DNA Evidence as Evidence in Criminal Trials

Familial Searching

Familial searching is the use of family members' DNA to identify a closely related suspect in jurisdictions where large DNA databases exist, but no exact match has been found. The first successful use of the practice was in a UK case where a man was convicted of manslaughter when he threw a brick stained with his own blood into a moving car.

Police could not get an exact match to the UK's DNA database because the man had no criminal convictions, but police implicated him using a close relative's DNA. The technique was used to catch a Los Angeles serial killer known as the "Grim Sleeper" in 2010.

However, critics have challenged the technology as "deeply antithetical to American values" and likely to lead to excess investigation of racial and ethnic minorities.

Surreptitious DNA Collecting

Police forces may collect DNA samples without the suspects' knowledge, and use it as evidence. Legality of this mode of proceeding has been questioned in Australia.

In the United States, it has been accepted, courts often claiming that there was no expectation of privacy, citing California v. Greenwood (1985), during which the Supreme Court held that the Fourth Amendment does not prohibit the warrantless search and seizure of garbage left for collection outside the curtilage of a home. Critics of this practice underline the fact that this analogy ignores that "most people have no idea that they risk surrendering their genetic identity to the police by, for instance, failing to destroy a used coffee cup.

Moreover, even if they do realize it, there is no way to avoid abandoning one's DNA in public." In the UK, the Human Tissue Act 2004 prohibited private individuals from covertly collecting biological samples (hair, fingernails, etc.) for DNA analysis, but excluded medical and criminal investigations from the offense.

England and Wales

Evidence from an expert who has compared DNA samples must be accompanied by evidence as to the sources of the samples and the procedures for obtaining the DNA profiles.

The judge must ensure that the jury must understand the significance of DNA matches and mismatches in the profiles. The judge must also ensure that the jury does not confuse the 'match probability' (the probability that a person that is chosen at random has a matching DNA profile to the sample from the scene) with the 'likelihood ratio' (the probability that a person with matching DNA committed the crime). In *R v. Doheny*, EWCA Crim 728. Phillips LJ gave this example of a summing up, which should be carefully tailored to the particular facts in each case:

Members of the Jury, if you accept the scientific evidence called by the Crown, this indicates that there are probably only four or five white males in the United Kingdom from whom that semen stain could have come.

The Defendant is one of them. If that is the position, the decision you have to reach, on all the evidence, is whether you are sure that it was the Defendant who left that stain or whether it is possible that it was one of that other small group of men who share the same DNA characteristics.

Juries should weigh up conflicting and corroborative evidence, using their own common sense and not by using mathematical formulae, such as Bayes' theorem, so as to avoid "confusion, misunderstanding and misjudgment".

Presentation and Evaluation of Evidence of Partial or Incomplete DNA Profiles

R v Bates (2006) EWCA Crim 1395 Moore-Bick LJ said: "We can see no reason why partial profile DNA evidence should not be admissible provided that the jury are made aware of its inherent limitations and are given a sufficient explanation to enable them to evaluate it.

There may be cases where the match probability in relation to all the samples tested is so great that the judge would consider its probative value to be minimal and decide to exclude the evidence in the exercise of his discretion, but this gives rise to no new question of principle and can be left for decision on a case by case basis.

However, the fact that there exists in the case of all partial profile evidence the possibility that a "missing" allele might exculpate the accused altogether does not provide sufficient grounds for rejecting such evidence. In many there is a possibility (at least in theory) that evidence exists which would assist the accused and perhaps even exculpate him altogether, but that does not provide grounds for excluding relevant evidence that is available and otherwise admissible, though it does make it important to ensure that the jury are given sufficient information to enable them to evaluate that evidence properly".

DNA Testing in the US

There are state laws on DNA profiling in all 50 states of the United States. Detailed information on database laws in each state can be found at the National Conference of State Legislatures website.

Development of Artificial DNA

In August 2009, scientists in Israel stunned the forensic sciences and raised serious questions concerning the use of DNA by law enforcement as the ultimate method of identification. In a paper published in the journal *Forensic Science International: Genetics*, the Israeli researchers demonstrated that it is possible to manufacture DNA in a laboratory, and thus falsify DNA evidence. The scientists fabricated saliva and blood samples, which originally contained DNA from a person other than the ostensible donor of the blood and saliva.

The researchers also showed that, using a DNA database, it is possible to take information from a profile and manufacture DNA to match it, and that this can be done without access to any actual DNA from the person whose DNA they are duplicating. The synthetic DNA oligos required for the procedure are common in molecular laboratories.

Dr. Daniel Frumkin, lead author on the paper, was quoted in *The New York Times* as saying, "You can just engineer a crime scene... any biology undergraduate could perform this."

Dr. Frumkin perfected a test that can forensically differentiate real DNA samples from fake ones. His test detects epigenetic modifications, in particular, DNA methylation. Seventy percent of the DNA in any human genome is methylated, meaning it contains methyl group modifications within a CpG dinucleotide context. Methylation at the promoter region is associated with gene silencing. The synthetic DNA lacks this epigenetic modification, which allows the test to distinguish manufactured DNA from original, genuine, DNA.

It is unknown how many, if any, police departments currently use the test, which appears to be a serious issue in light of Frumkin's claim that the DNA manufacturing procedure is within the grasp of any undergraduate biology student. No police lab has publicly announced that it is using the new test to verify DNA results, while FSI Genetics says that any forensic laboratory doing DNA identification should adopt this test to authenticate its results as "real" DNA.

Bioinformatics

Bioinformatics involves the manipulation, searching, and data mining of DNA sequence data. The development of techniques to store and search DNA sequences have led to widely applied advances in computer science, especially string searching algorithms, machine learning and database theory. String searching or matching algorithms, which find an occurrence of a sequence of letters inside a larger sequence of letters, were developed to search for specific sequences of nucleotides. In other applications such as text editors, even simple algorithms for this problem usually suffice, but DNA sequences cause these algorithms to exhibit near-worst-case behaviour due to their small number of distinct characters. The related problem of sequence alignment aims to identify homologous sequences and locate the specific mutations that make them distinct. These techniques, especially multiple sequence alignment, are used in studying phylogenetic relationships and protein function.

Data sets representing entire genomes' worth of DNA sequences, such as those produced by the Human Genome Project, are difficult to use without annotations, which label the locations of genes and regulatory elements on each chromosome. Regions of DNA sequence that have the characteristic patterns associated with protein- or RNA-coding genes can be identified by gene finding algorithms, which allow researchers to predict the presence of particular gene products in an organism even before they have been isolated experimentally.

DNA Nanotechnology

DNA nanotechnology uses the unique molecular recognition properties of DNA and other nucleic acids to create self-assembling branched DNA complexes with useful properties. DNA is thus used as a structural material rather than as a carrier of biological information. This has led to the creation of two-dimensional periodic lattices (both tile-based as well as using the "DNA origami" method) as well as three-dimensional structures in the shapes of polyhedra. Nanomechanical devices and algorithmic self-assembly have also been demonstrated, and these DNA structures have been used to template the arrangement of other molecules such as gold nanoparticles and streptavidin proteins.

History and Anthropology

Because DNA collects mutations over time, which are then inherited, it contains historical information, and, by comparing DNA sequences, geneticists can infer the evolutionary history of organisms, their phylogeny. This field of phylogenetics is a powerful tool in evolutionary biology. If DNA sequences within a species are compared, population geneticists can learn the history of particular populations. This can be used in studies ranging from ecological genetics to anthropology; For example, DNA evidence is being used to try to identify the Ten Lost Tribes of Israel.

DNA has also been used to look at modern family relationships, such as establishing family relationships between the descendants of Sally Hemings and Thomas Jefferson. This usage is closely related to the use of DNA in criminal investigations detailed above. Indeed, some criminal investigations have been solved when DNA from crime scenes has matched relatives of the guilty individual.

History of DNA Research

DNA was first isolated by the Swiss physician Friedrich Miescher who, in 1869, discovered a microscopic substance in the pus of discarded surgical bandages. As it resided in the nuclei of cells, he called it "nuclein". In 1919, Phoebus Levene identified the base, sugar and phosphate nucleotide unit. Levene suggested that DNA consisted of a string of nucleotide units linked together through the phosphate groups. However, Levene thought the chain was short and the bases repeated in a fixed order. In 1937 William Astbury produced the first X-ray diffraction patterns that showed that DNA had a regular structure.

In 1928, Frederick Griffith discovered that traits of the "smooth" form of the *Pneumococcus* could be transferred to the "rough" form of the same bacteria by mixing killed "smooth" bacteria with the live "rough" form. This system provided the first clear suggestion that DNA carries genetic information—the Avery- MacLeod-McCart experiment—when Oswald Avery, along with coworkers Colin MacLeod and Maclyn McCarty, identified DNA as the transforming principle in 1943. DNA's role in heredity was confirmed in 1952, when Alfred Hershey and Martha Chase in the Hershey-Chase experiment showed that DNA is the genetic material of the T2 phage. In 1953, James D. Watson and Francis Crick suggested what is now accepted as the first correct double-helix model of DNA structure in the journal *Nature*. Their double-helix, molecular model of DNA was then based on a single X-ray diffraction image (labeled as "Photo 51") taken by Rosalind Franklin and Raymond Gosling in May 1952, as well as the information that the DNA bases are paired — also obtained through private communications from Erwin Chargaff in the previous years. Chargaff's rules played a very important role in establishing double-helix configurations for B-DNA as well as A-DNA.

Experimental evidence supporting the Watson and Crick model were published in a series of five articles in the same issue of *Nature*. Of these, Franklin and Gosling's paper was the first publication of their own X-ray diffraction data and original analysis method that partially supported the Watson and Crick model; this issue also contained an article on DNA structure by Maurice Wilkins and two of his colleagues, whose analysis and *in vivo* B-DNA X-ray patterns also supported the presence *in vivo* of the double-helical DNA configurations as proposed by Crick and Watson for their double-helix molecular model of DNA in the previous two pages of *Nature*. In 1962, after Franklin's death, Watson, Crick, and Wilkins jointly received the Nobel Prize in Physiology or Medicine. However, Nobel rules of the time allowed only living recipients, but a vigorous debate continues on who should receive credit for the discovery.

In an influential presentation in 1957, Crick laid out the "Central Dogma" of molecular biology, which foretold the relationship between DNA, RNA, and proteins, and articulated the "adaptor hypothesis". Final confirmation of the replication mechanism that was implied by the double-helical structure followed in 1958 through the Meselson-Stahl experiment. Further work by Crick and coworkers showed that the genetic code was based on non-overlapping triplets of bases, called codons, allowing Har Gobind Khorana, Robert W. Holley and Marshall Warren Nirenberg to decipher the genetic code. These findings represent the birth of molecular biology.

Protein Synthesis

Protein synthesis is the process in which cells build proteins. The term is sometimes used to refer only to protein translation but more often it refers to a multi-step process, beginning with amino acid synthesis and transcription of nuclear DNA into messenger RNA, which is then used as input to translation.

The cistron DNA is transcribed into a variety of RNA intermediates. The last version is used as a template in synthesis of a polypeptide chain. Proteins can often be synthesized directly from genes by translating mRNA. When a protein needs to be available on short notice or in large quantities, a protein precursor is produced. A proprotein is an inactive protein containing one or more inhibitory peptides that can be activated when the inhibitory sequence is removed by proteolysis during posttranslational modification.

A preprotein is a form that contains a signal sequence (an N-terminal signal peptide) that specifies its insertion into or through membranes; i.e., targets them for secretion. The signal peptide is cleaved off in the endoplasmic reticulum. Preproproteins have both sequences (inhibitory and signal) still present. For synthesis of protein, a succession of tRNA molecules charged with appropriate amino acids have to be brought together with an mRNA molecule and matched up by base-pairing through their anti-codons with each of its successive codons.

The amino acids then have to be linked together to extend the growing protein chain, and the tRNAs, relieved of their burdens, have to be released. This whole complex of processes is carried out by a giant multimolecular machine, the ribosome, formed of two main chains of RNA, called ribosomal RNA (rRNA), and more than 50 different proteins.

This molecular juggernaut latches onto the end of an mRNA molecule and then trundles along it, capturing loaded tRNA molecules and stitching together the amino acids they carry to form a new protein chain. Protein biosynthesis, although very similar, is different for prokaryotes and eukaryotes.

Amino Acid Synthesis

Amino acid synthesis is the set of biochemical processes (metabolic pathways) by which the various amino acids are produced from other compounds. The substrates for these processes are various compounds in the organism's diet or growth media. Not all organisms are able to synthesise all amino acids, for example humans are only able to synthesise 12 of the 20 standard amino acids.

A fundamental problem for biological systems is to obtain nitrogen in an easily usable form. This problem is solved by certain microorganisms capable of reducing the inert Na"N molecule (nitrogen gas) to two molecules of ammonia in one of the most remarkable reactions in biochemistry.

Nitrogen in the form of ammonia is the source of nitrogen for all the amino acids. The carbon backbones come from the glycolytic pathway, the pentose phosphate pathway, or the citric acid cycle.

In amino acid production, one encounters an important problem in biosynthesis— namely, stereochemical control. Because all amino acids except glycine are chiral, biosynthetic pathways must generate the correct isomer with high fidelity. In each of the 19 pathways for the generation of chiral amino acids, the stereochemistry at the á-carbon atom is established by a transamination reaction that involves pyridoxal phosphate. Almost all the transaminases that catalyze these reactions descend from a common ancestor, illustrating once again that effective solutions to biochemical problems are retained throughout evolution.

Biosynthetic pathways are often highly regulated such that building blocks are synthesized only when supplies are low. Very often, a high concentration of the final product of a pathway inhibits the activity of enzymes that function early in the pathway. Often present are allosteric enzymes capable of sensing and responding to concentrations of regulatory species. These enzymes are similar in functional properties to aspartate transcarbamoylase and its regulators. Feedback and allosteric mechanisms ensure that all twenty amino acids are maintained in sufficient amounts for protein synthesis and other processes.

Amino acids are synthesized from α-ketoacids, and later transaminated from another aminoacid, usually Glutamate. The enzyme involved in this reaction is an aminotransferase.

Glutamate itself is formed by amination of α-ketoglutarate:

Nitrogen Fixation: Microorganisms Use ATP and a Powerful Reductant to Reduce Atmospheric Nitrogen to Ammonia

Microorganisms use ATP and reduced ferredoxin, a powerful reductant, to reduce N2 to NH3. An iron-molybdenum cluster in nitrogenase deftly catalyzes the fixation of N2, a very inert molecule. Higher organisms consume the fixed nitrogen to synthesize amino acids, nucleotides, and other nitrogen-containing biomolecules. The major points of entry of NH4+ into metabolism are glutamine or glutamate.

Amino Acids are Made from Intermediates of the Citric Acid Cycle and Other Major Pathways

Of the basic set of 20 amino acids (not counting selenocysteine) there are 8 that human beings cannot synthesize. In addition, the amino acids arginine, cysteine, glycine, glutamine, histidine, proline, serine and tyrosine are considered conditionally essential, meaning they are not normally required in the diet, but must be supplied exogenously to specific populations that do not synthesize it in adequate amounts. (For example, enough arginine is synthesized by the urea cycle to meet the needs of an adult but perhaps not those of a growing child.)

Amino acids that need to be obtained from the diet are called essential amino acids. Nonessential amino acids are produced in the body. The pathways for the synthesis of nonessential amino acids are quite simple. Glutamate dehydrogenase catalyzes the reductive amination of α-ketoglutarate to glutamate. A transamination reaction takes place in the synthesis of most amino acids. At this step, the chirality of the amino acid is established. Alanine and aspartate are synthesized by the transamination of pyruvate and oxaloacetate, respectively. Glutamine is synthesized from NH4+ and glutamate, and asparagine is synthesized similarly. Proline and arginine are derived from glutamate.

Serine, formed from 3-phosphoglycerate, is the precursor of glycine and cysteine. Tyrosine is synthesized by the hydroxylation of phenylalanine, an essential amino acid. The pathways for the biosynthesis of essential amino acids are much more complex than those for the nonessential ones. Tetrahydrofolate, a carrier of activated one-carbon units, plays an important role in the metabolism of amino acids and nucleotides. This coenzyme carries one-carbon units at three oxidation states, which are interconvertible: most reduced—methyl; intermediate—methylene; and most oxidized—formyl, formimino, and methenyl. The major donor of activated methyl groups is S-adenosylmethionine, which is synthesized by the transfer of an adenosyl group from ATP to the sulfur atom of methionine. S-Adenosylhomocysteine is formed when the activated methyl group is transferred to an acceptor. It is hydrolyzed to adenosine and homocysteine, the latter of which is then methylated to methionine to complete the activated methyl cycle. The A Cortisol inhibites protein synthesis.

Amino Acid Biosynthesis is Regulated by Feedback Inhibition

Most of the pathways of amino acid biosynthesis are regulated by feedback inhibition, in which the committed step is allosterically inhibited by the final product. Branched pathways require extensive interaction among the branches that includes both negative and positive regulation. The regulation of glutamine synthetase from E. coli is a striking demonstration of cumulative feedback inhibition and of control by a cascade of reversible covalent modifications.

Amino Acids are Precursors of Many Biomolecules

Amino acids are precursors of a variety of biomolecules. Glutathione serves as a sulfhydryl buffer and detoxifying agent. Glutathione peroxidase, a selenoenzyme, catalyzes the reduction of hydrogen peroxide and organic peroxides by glutathione. Nitric oxide, a short-lived messenger, is formed from arginine. Porphyrins are synthesized from glycine and succinyl CoA, which condense to give β-aminolevulinate. Two molecules of this intermediate become linked to form porphobilinogen. Four molecules of porphobilinogen combine to form a linear tetrapyrrole, which cyclizes to uroporphyrinogen III. Oxidation and side-chain modifications lead to the synthesis of protoporphyrin IX, which acquires an iron atom to form heme.

Transcription

In transcription an mRNA chain is generated, with one strand of the DNA double helix in the genome as template. This strand is called the template strand. Transcription can be divided into 3 stages: Initiation, Elongation and Termination, each regulated by a large number of proteins such as transcription factors and coactivators that ensure the correct gene is transcribed.

The DNA strand is read in the 3' to 5' direction and the mRNA is transcribed in the 5' to 3' direction by the RNA polymerase. Transcription occurs in the cell nucleus, where the DNA is held. The DNA structure of the cell is made up of two helixes made up of sugar and phosphate held together by the bases. The sugar and the phosphate are joined together by covalent bond.

The DNA is "unzipped" by the enzyme helicase, leaving the single nucleotide chain open to be copied. RNA polymerase reads the DNA strand from 3 prime (3') end to the 5 prime (5') end, while it synthesizes a single strand of messenger RNA in the 5' to 3' direction. The general RNA structure is very similar to the DNA structure, but in RNA the nucleotide uracil takes the place that thymine occupies in DNA. The single strand of mRNA leaves the nucleus through nuclear pores, and migrates into the cytoplasm.

The first product of transcription differs in prokaryotic cells from that of eukaryotic cells, as in prokaryotic cells the product is mRNA, which needs no post-transcriptional modification, while in eukaryotic cells, the first product is called primary transcript, that needs post-transcriptional modification (capping with 7 methyl guanosine, tailing with a poly A tail) to give hnRNA (heterophil nuclear RNA). hnRNA then undergoes splicing of introns (non coding parts of the gene) via spliceosomes to produce the final mRNA.

Translation

The synthesis of proteins is known as translation. Translation occurs in the cytoplasm, where the ribosomes are located. Ribosomes are made of a small and large subunit that surround the mRNA. In translation, messenger RNA (mRNA) is decoded to produce a specific polypeptide according to the rules specified by the trinucleotide genetic code. This uses an mRNA sequence as a template to guide the synthesis of a chain of amino acids that form a protein. Translation proceeds in four phases: activation, initiation, elongation, and termination (all describing the growth of the amino acid chain, or polypeptide that is the product of translation).

In activation, the correct amino acid (AA) is joined to the correct transfer RNA (tRNA). While this is not technically a step in translation, it is required for translation to proceed. The AA is joined by its carboxyl group to the 3' OH of the tRNA by an ester bond.

When the tRNA has an amino acid linked to it, it is termed "charged". Initiation involves the small subunit of the ribosome binding to 5' end of mRNA with the help of initiation factors (IF), other proteins that assist the process. Elongation occurs when the next aminoacyl-tRNA (charged tRNA) in line binds to the ribosome along with GTP and an elongation factor.

Termination of the polypeptide happens when the A site of the ribosome faces a stop codon (UAA, UAG, or UGA). When this happens, no tRNA can recognize it, but releasing factor can recognize nonsense codons and causes the release of the polypeptide chain. The capacity of disabling or inhibiting translation in protein biosynthesis is used by antibiotics such as: anisomycin, cycloheximide, chloramphenicol, tetracycline, streptomycin, erythromycin, puromycin etc.

Translation is the process of converting the mRNA codon sequences into an amino acid polypeptide chain.

1. Amino acid activation
2. Initiation - A ribosome attaches to the mRNA and starts to code at the FMet codon (usually AUG, sometimes GUG or UUG).
3. Elongation - tRNA brings the corresponding amino acid (which has an anticodon that identifies the amino acid as the corresponding molecule to a codon) to each codon as the ribosome moves down the mRNA strand.
4. Termination - Reading of the final mRNA codon (aka the STOP codon), which ends the synthesis of the peptide chain and releases it.

Events Following Protein Translation

The events following biosynthesis include post-translational modification and protein folding. During and after synthesis, polypeptide chains often fold to assume, so called, native secondary and tertiary structures. This is known as *protein folding*.

Many proteins undergo *post-translational modification*. This may include the formation of disulfide bridges or attachment of any of a number of biochemical functional groups, such as acetate, phosphate, various lipids and carbohydrates. Enzymes may also remove one or more amino acids from the leading (amino) end of the polypeptide chain, leaving a protein consisting of two polypeptide chains connected by disulfide bonds.

In general, protein molecules are believed to be modified by small chemical groups, post-translationally. Chemical modifications such as, phosphorylation of serine/threonine, acetylation or methylation of lysine, hydroxylation of proline/lysine, formylation of glycine, glycosylation of serine/threonine/ asparagine, acylation of cysteine, myristoylation of glycine, biotinylation of lysine, ubiquitination, etc. on proteins is a very important issue in relation to properly understanding the biological functions of a given protein. These much-studied post-translational modifications have become well-established with the discovery of the respective enzymes (kinases for phosphorylation, acetylases for acetylation, methyl transferases for methylation, etc.), which carry on the chemical modifications on the specific amino acid residues.

All of these modifications are still believed to have happened as post-translational events. There is no study yet on when actually one particular modification occurs on a given amino acid residue in a given protein. Does it happen when the protein is already formed, or when the amino acid chain is being synthesized, or before the translation of the primary chain has begun?

Since these chemical modifications are related to the biological functions of a protein, it is easy to think that these chemical modifications have happened to the whole protein molecule, after the protein primary chain is fully synthesized; but, if that is the case, we have to consider the fact that the primary chains get folded instantly, (in a similar way as the newly synthesized DNA strands form helixes), to attain its compact-globular conformation ; As most of the primary chains are fairly long [a 5Kd protein may have 40-45 amino acid residues in its primary chain], it is likely that the newly formed amino acid chain tries to remain intact by folding, thereby avoiding its breakdown via lots of proteases present within the cytoplasm. And, no capping event to protect the N-terminal end of the primary sequence (similar to 5' m-RNA capping to protect m-RNAs) is ever discovered for protein primary structure. So, by folding mechanism, the primary chain, perhaps, avoids the protease attacks.

However, once it gets folded, it may be very difficult for the respective enzyme molecule to find out the particular aa residue from the complexity of that compactly folded conformation. In addition, it can be clearly imagined that this enzymatic modification/reaction on a given amino acid requires presence and association of the appropriate enzyme, necessary cofactors, etc. This association is much easier to occur when the amino acid residues in the primary structure are readily available for binding; in other words, it is much difficult for the enzyme molecules to find and to bind to its substrate amino acid residue in a mature protein molecule after its three-dimensional conformation been attained.

So one can think that the modifications can happen while the primary chain of the protein is being synthesized during the translation process on the m-RNA strand; the amino acid residues on the primary chain can be modified instantly and enzymatically by kinases, acetylases, hydroxylases, methyltransferases, etc. to initiate proper folding for protection in order to avoid degradation by proteases, thereby gaining the globular form.

Also, the reader can imagine another scenario in which, while the free amino acid molecules are formed within a cell and become available, they [in that free state] may be modified enzymatically before taking the ride to the translational event; this means that, while the primary chain is being synthesized, the pre-modified amino acid molecules are ready to be engaged.

This way, as the primary chain is getting synthesized, it does not have to be modified by the modifying enzymes anymore, and can fold itself instantaneously without concern of degradation. This also arises new thoughts that (i) phosphate, acetyl, methyl, biotinyl, acyl, etc. groups may have the ability to inhibit protease actions on the primary chains; (ii) most or all of the amino acid residues get modified by small chemical groups (so far, only some chemical groups are known).

It is still not fully known exactly when and how the actual modification of a given amino acid residue occurs, at which stage of synthesis, within a protein molecule.

Once the chemically-modified, protease-insensitive, intact protein molecule is generated, it must perform its biological function, which requires its being activated. The activation is probably done by a second set of enzymatic reactions when these chemical groups are removed from the a residues (or added back). So, the second type of post-translational modifications are the opposite reactions of the above-described type, which are dephosphorylation by phosphatases or phosphoryl transferases, deacetylation by deacetylases or acetyltransferases, demethylation by demethylases or methyl-transferases, ubiquitination, SuMoylation, glycosylation, biotinylation, etc.

These are taking place on the whole protein molecule toward generating its activated form to execute a particular function or toward its deactivation followed by its total degradation. As the chemical groups are removed (or added), it is quite clear that the proteins can go through different states of structural/conformational change. Thus, after translation, a protein can change its conformation dynamically while the chemical groups are removed or added enzymatically; these conformational changes in the protein structure help the protein to proceed through its lifecycle, until it is ubiquitinated for its total degradation.

5

Genetic Engineering

Genome organization refers to the sequential, not the structural organization of the genome. Besides the coding exons, the non-coding DNA in Eukaryotes may fall in the following classes:

- Introns. They are DNA sequences inserted between the exons and found in the ORF. They are spliced after the first level of transcription. Most introns are junk inserted within genes.
- Pseudogenes. 'Dead', non-functional copies of genes present elsewhere in the genome, but no longer of any use.
- Retropseudogenes. Like pseudogenes, but have been processed, i.e. lack introns. Produced by the action of reverse transcriptase (RT) on mRNA, and subsequent incorporation of the cDNA into the genome.
- Transposons. Jumping genes, which splice themselves in and out of the genome (in DNA form) randomly, by the action of transposase.
- Retrotransposons. Transcribed into an mRNA, which encodes an RT enzyme, which then copies the mRNA back to DNA and incorporates it into the genome.

Infact in humans only 1.5% of the entire genome length corresponds to coding DNA. This 1.5% codes for about 27,000 genes which in turn code for proteins that are responsible for all the cellular processes.

Genome Analysis?

Genome analysis entails the prediction of genes in uncharacterized genomic sequences. The 21st century has seen the announcement of the draft version of the human genome sequence. Model organisms have been sequenced in both the plant and animal kingdoms. However, the pace of genome annotation is not matching the pace of genome sequencing. Experimental genome annotation is slow and time consuming. The demand is to be able to develop computational tools for gene prediction.

Computational Gene prediction is relatively simple for the prokaryotes where all the genes are converted into the corresponding mRNA and then into proteins. The process is more complex for eukaryotic cells where the coding DNA sequence is interrupted by random sequences called introns.

Some of the questions which biologists want to answer today are:

- Given a DNA sequence, what part of it codes for a protein and what part of it is junk DNA.
- Classify the junk DNA as intron, untranslated region, transposons, dead genes, regulatory elements etc.
- Divide a newly sequenced genome into the genes (coding) and the non-coding regions.

The importance of genome analysis can be understood by comparing the human and chimpanzee genomes. The chimp and human genomes vary by an average of just 2% i.e. just about 160 enzymes. A complete genome analysis of the two genomes would give a strong insight into the various mechanisms responsible for the differences. Arabidopsis and Humans have the same number of genes, though the Arabidopsis genome is around 250 times smaller than humans. How is that?

The human genome has a lot of junk DNA, specifically transposons and mobile genetic elements. This increases the size of the human genome, though the number of genes is only 27,000. However, the number of protein products in humans is significantly higher. Many of the sequenced human genes have alternative splice products. In addition, several other processes (e.g. signal transduction) proceed via further protein modifications, such as Glycosylation. Therefore, the number of human protein products could far exceed the number of genes.

Why do plants have such bulky genomes when they are not as complex as some of the higher eukaryotes ?

This is mostly due to two factors: the ability of plants to duplicate their genomes in order to reproduce (a process known as polyploidization) and the susceptibility of plants to mobile genetic elements.

- Polyploidization allows plants to more easily form hybrids when pollen and ova from different species come together. The result of such hybridization events are plants with genomes that are the sum of the two parent genome sizes (as opposed to half of one parent's genome and half of the other parent's genome as in normal sexual reproduction.

- Also, in case of plants, it is fairly common to observe insertion of transposable elements in intergenic regions. This also explains the difference in the sizes of plant genomes among themselves as well.

Open Reading Frame

The region of the nucleotide sequences from the start codon (ATG) to the stop codon is called the Open Reading frame. Gene finding in organism specially prokaryotes starts form searching for an open reading frames (ORF). An ORF is a sequence of DNA that starts with start codon "ATG" (not always) and ends with any of the three termination codons (TAA, TAG, TGA). Depending on the starting point, there are six possible ways (three on forward strand and three on complementary strand) of translating any nucleotide sequence into amino acid sequence according to the genetic code.These are called reading frames. While eukaryotic gene finding is altogether a different task as the eukaryotic genes are not continuous and interrupted by intervening noncoding sequences called 'introns'. Moreover organization of genetic information in eukaryotes and prokaryotes is different.

What is Coding Sequence

The Coding Sequence (CDS) is the actual region of DNA that is translated to form proteins. While the ORF may contain introns as well, the CDS refers to those nucleotides(concatenated exons) that can be divided into codons which are actually translated into amino acids by the ribosomal translation machinery. In Prokaryotes the ORF and the CDS are the same.

DNA Replication

Before a cell divides, its DNA is replicated (duplicated.) Because the two strands of a DNA molecule have complementary base pairs, the nucleotide sequence of each strand automatically supplies the information needed to produce its partner.

If the two strands of a DNA molecule are separated, each can be used as a pattern or template to produce a complementary strand. Each template and its new complement together then form a new DNA double helix, identical to the original. Before replication can occur, the length of the DNA double helix about to be copied must be unwound.

In addition, the two strands must be separated, much like the two sides of a zipper, by breaking the weak hydrogen bonds that link the paired bases. Once the DNA strands have been unwound, they must be held apart to expose the bases so that new nucleotide partners can hydrogen-bond to them.

The enzyme DNA polymerase then moves along the exposed DNA strand, joining newly arrived nucleotides into a new DNA strand that is complementary to the template. Each cell contains a family of more than thirty enzymes to insure the accurate replication of DNA.

Primers

Though DNA polymerase can elongate a polynucleotide strand by adding new nucleotides, it cannot start a strand from scratch because it can only bond new nucleotides to a free sugar (3') end of a nucleotide chain. DNA polymerase requires the assistance of a primer, a previously existing short strand of DNA (or RNA) that is complementary to the first part of the DNA segment being copied. This small strand of nucleotides anneals (binds) by complementary base pairing to the beginning of the area being copied. With the primer in place, DNA polymerase is then able to continue adding the rest of the pairs of the segment until a new double strand of DNA is completed. Primers are formed from free nucleotides in the cell by enzymes called DNA primases.

Replication Occurs Differently on Antiparallel Strands of DNA

That nucleotides can be added only to the sugar or 3' end of the growing complementary chain presents no problem for the side of the DNA chain opening at its phosphate or 5' end. The primer that binds to the first few exposed bases will end with a sugar (3') where the phosphate of a new nucleotide can be attached. From there on, DNA polymerase can continuously synthesize the growing complementary strand. This strand of DNA is called the leading strand. A different challenge faces DNA polymerase when the complementary side of the DNA molecule begins unzipping from its sugar (3') toward its phosphate (5') end. A primer of complementary molecules attaching to the opening end of this chain would have a phosphate not a sugar at its exposed end so that new nucleotides could not be joined.

To get around this problem, this strand is synthesized in small pieces backward from the overall direction of replication. This strand is called the lagging strand. The short segments of newly assembled DNA from which the lagging strand is built are called Okazaki fragments. As replication proceeds and nucleotides are added to the 3' end of the Okazaki fragments, they come to meet each other. The primer fragments are then booted out by enzymes and replaced by appropriate DNA nucleotides. The whole thing is then stitched together by another enzyme called DNA ligase.

DNA Damage, Repair, and Recombination

DNA repair refers to a collection of processes by which a cell identifies and corrects damage to the DNA molecules that encode its genome. In human cells, both normal metabolic activities and environmental factors such as UV light and radiation can cause DNA damage, resulting in as many as 1 million individual molecular lesions per cell per day. Many of these lesions cause structural damage to the DNA molecule and can alter or eliminate the cell's ability to transcribe the gene that the affected DNA encodes.

Other lesions induce potentially harmful mutations in the cell's genome, which affect the survival of its daughter cells after it undergoes mitosis. As a consequence, the DNA repair process is constantly active as it responds to damage in the DNA structure. When normal repair processes fail, and when cellular apoptosis does not occur, irreparable DNA damage may occur, including double-strand breaks and DNA crosslinkages.

The rate of DNA repair is dependent on many factors, including the cell type, the age of the cell, and the extracellular environment. A cell that has accumulated a large amount of DNA damage, or one that no longer effectively repairs damage incurred to its DNA, can enter one of three possible states:

1. an irreversible state of dormancy, known as senescence
2. cell suicide, also known as apoptosis or programmed cell death
3. unregulated cell division, which can lead to the formation of a tumor that is cancerous.

The DNA repair ability of a cell is vital to the integrity of its genome and thus to its normal functioning and that of the organism. Many genes that were initially shown to influence life span have turned out to be involved in DNA damage repair and protection. Failure to correct molecular lesions in cells that form gametes can introduce mutations into the genomes of the offspring and thus influence the rate of evolution.

DNA Damage

DNA damage, due to environmental factors and normal metabolic processes inside the cell, occurs at a rate of 1,000 to 1,000,000 molecular lesions per cell per day. While this constitutes only 0.000165% of the human genome's approximately 6 billion bases (3 billion base pairs), unrepaired lesions in critical genes (such as tumor suppressor genes) can impede a cell's ability to carry out its function and appreciably increase the likelihood of tumour formation.

The vast majority of DNA damage affects the primary structure of the double helix; that is, the bases themselves are chemically modified. These modifications can in turn disrupt the molecules' regular helical structure by introducing non-native chemical bonds or bulky adducts that do not fit in the standard double helix. Unlike proteins and RNA, DNA usually lacks tertiary structure and therefore damage or disturbance does not occur at that level.

DNA is, however, supercoiled and wound around "packaging" proteins called histones (in eukaryotes), and both superstructures are vulnerable to the effects of DNA damage.

Sources of Damage

DNA damage can be subdivided into two main types:

1. endogenous damage such as attack by reactive oxygen species produced from normal metabolic byproducts (spontaneous mutation), especially the process of oxidative deamination
   - also includes replication errors.
2. exogenous damage caused by external agents such as
   - ultraviolet [UV 200-300nm] radiation from the sun
   - other radiation frequencies, including x-rays and gamma rays
   - hydrolysis or thermal disruption
   - certain plant toxins
   - human-made mutagenic chemicals, especially aromatic compounds that act as DNA intercalating agents
   - cancer chemotherapy and radiotherapy
   - viruses.

The replication of damaged DNA before cell division can lead to the incorporation of wrong bases opposite damaged ones. Daughter cells that inherit these wrong bases carry mutations from which the original DNA sequence is unrecoverable (except in the rare case of a back mutation, for example, through gene conversion).

Types of Damage

There are five main types of damage to DNA due to endogenous cellular processes:

1. *oxidation* of bases [e.g. 8-oxo-7,8-dihydroguanine (8-oxoG)] and generation of DNA strand interruptions from reactive oxygen species,
2. *alkylation* of bases (usually methylation), such as formation of 7-methylguanine, 1-methyladenine, 6-O-Methylguanine
3. *hydrolysis* of bases, such as deamination, depurination, and depyrimidination.
4. "bulky adduct formation" (i.e., benzo[a]pyrene diol epoxide-dG adduct)
5. *mismatch* of bases, due to errors in DNA replication, in which the wrong DNA base is stitched into place in a newly forming DNA strand, or a DNA base is skipped over or mistakenly inserted.

Damage caused by exogenous agents comes in many forms. Some examples are:

1. *UV-B light* causes crosslinking between adjacent cytosine and thymine bases creating *pyrimidine dimers*. This is called direct DNA damage.
2. *UV-A light* creates mostly free radicals. The damage caused by free radicals is called indirect DNA damage.
3. *Ionizing radiation* such as that created by radioactive decay or in *cosmic rays* causes breaks in DNA strands. Low-level ionizing radiation may induce irreparable DNA damage (leading to replicational and transcriptional errors needed for neoplasia or may trigger viral interactions) leading to pre-mature aging and cancer.
4. *Thermal disruption* at elevated temperature increases the rate of depurination (loss of purine bases from the DNA backbone) and single-strand breaks. For example, hydrolytic depurination is seen in the thermophilic bacteria, which grow in hot springs at 40-80 °C. The rate of depurination (300 purine residues per genome per generation) is too high in these species to be repaired by normal repair machinery, hence a possibility of an adaptive response cannot be ruled out.
5. *Industrial chemicals* such as vinyl chloride and hydrogen peroxide, and environmental chemicals such as polycyclic hydrocarbons found in smoke, soot and tar create a huge diversity of DNA adducts- ethenobases, oxidized bases, alkylated phosphotriesters and Crosslinking of DNA just to name a few.

UV damage, alkylation/methylation, X-ray damage and oxidative damage are examples of induced damage. Spontaneous damage can include the loss of a base, deamination, sugar ring puckering and tautomeric shift.

Nuclear versus Mitochondrial DNA Damage

In human cells, and eukaryotic cells in general, DNA is found in two cellular locations - inside the nucleus and inside the mitochondria. Nuclear DNA (nDNA) exists as chromatin during non-replicative stages of the cell cycle and is condensed into aggregate structures known as chromosomes during cell division. In either state the DNA is highly compacted and wound up around bead-like proteins called histones.

Whenever a cell needs to express the genetic information encoded in its nDNA the required chromosomal region is unravelled, genes located therein are expressed, and then the region is condensed back to its resting conformation. Mitochondrial DNA (mtDNA) is located inside mitochondria organelles, exists in multiple copies, and is also tightly associated with a number of proteins to form a complex known as the nucleoid. Inside mitochondria, reactive oxygen species (ROS), or free radicals, byproducts of the constant production of adenosine triphosphate (ATP) via oxidative phosphorylation, create a highly oxidative environment that is known to damage mtDNA. A critical enzyme in counteracting the toxicity of these species is superoxide dismutase, which is present in both the mitochondria and cytoplasm of eukaryotic cells.

Senescence and Apoptosis

Senescence, an irreversible state in which the cell no longer divides, is a protective response to the shortening of the chromosome ends. The telomeres are long regions of repetitive noncoding DNA that cap chromosomes and undergo partial degradation each time a cell undergoes division. In contrast, quiescence is a reversible state of cellular dormancy that is unrelated to genome damage.

Senescence in cells may serve as a functional alternative to apoptosis in cases where the physical presence of a cell for spatial reasons is required by the organism, which serves as a "last resort" mechanism to prevent a cell with damaged DNA from replicating inappropriately in the absence of pro-growth cellular signaling. Unregulated cell division can lead to the formation of a tumour, which is potentially lethal to an organism. Therefore, the induction of senescence and apoptosis is considered to be part of a strategy of protection against cancer.

DNA Damage and Mutation

It is important to distinguish between DNA damage and mutation, the two major types of error in DNA. DNA damages and mutation are fundamentally different. Damages are physical abnormalities in the DNA, such as single- and double-strand breaks, 8-hydroxydeoxyguanosine residues, and polycyclic aromatic hydrocarbon adducts. DNA damages can be recognized by enzymes, and, thus, they can be correctly repaired if redundant information, such as the undamaged sequence in the complementary DNA strand or in a homologous chromosome, is available for copying. If a cell retains DNA damage, transcription of a gene can be prevented, and, thus, translation into a protein will also be blocked. Replication may also be blocked and/or the cell may die.

In contrast to DNA damage, a mutation is a change in the base sequence of the DNA. A mutation cannot be recognized by enzymes once the base change is present in both DNA strands, and, thus, a mutation cannot be repaired. At the cellular level, mutations can cause alterations in protein function and regulation. Mutations are replicated when the cell replicates. In a population of cells, mutant cells will increase or decrease in frequency according to the effects of the mutation on the ability of the cell to survive and reproduce. Although distinctly different from each other, DNA damages and mutations are related because DNA damages often cause errors of DNA synthesis during replication or repair; these errors are a major source of mutation. Given these properties of DNA damage and mutation, it can be seen that DNA damages are a special problem in non-dividing or slowly dividing cells, where unrepaired damages will tend to accumulate over time. On the other hand, in rapidly dividing cells, unrepaired DNA damages that do not kill the cell by blocking replication will tend to cause replication errors and thus mutation. The great majority of mutations that are not neutral in their effect are deleterious to a cell's survival.

Thus, in a population of cells comprising a tissue with replicating cells, mutant cells will tend to be lost. However, infrequent mutations that provide a survival advantage will tend to clonally expand at the expense of neighboring cells in the tissue. This advantage to the cell is disadvantageous to the whole organism, because such mutant cells can give rise to cancer. Thus, DNA damages in frequently dividing cells, because they give rise to mutations, are a prominent cause of cancer. In contrast, DNA damages in infrequently dividing cells are likely a prominent cause of aging.

DNA Repair Mechanisms

Cells cannot function if DNA damage corrupts the integrity and accessibility of essential information in the genome (but cells remain superficially functional when so-called "non-essential" genes are missing or damaged). Depending on the type of damage inflicted on the DNA's double helical structure, a variety of repair strategies have evolved to restore lost information. If possible, cells use the unmodified complementary strand of the DNA or the sister chromatid as a template to recover the original information. Without access to a template, cells use an error-prone recovery mechanism known as translesion synthesis as a last resort. Damage to DNA alters the spatial configuration of the helix, and such alterations can be detected by the cell. Once damage is localized, specific DNA repair molecules bind at or near the site of damage, inducing other molecules to bind and form a complex that enables the actual repair to take place.

Direct Reversal

Cells are known to eliminate three types of damage to their DNA by chemically reversing it. These mechanisms do not require a template, since the types of damage they counteract can occur in only one of the four bases. Such direct reversal mechanisms are specific to the type of damage incurred and do not involve breakage of the phosphodiester backbone.

The formation of pyrimidine dimers upon irradiation with UV light results in an abnormal covalent bond between adjacent pyrimidine bases. The photoreactivation process directly reverses this damage by the action of the enzyme photolyase, whose activation is obligately dependent on energy absorbed from blue/UV light (300–500 nm wavelength) to promote catalysis. Another type of damage, methylation of guanine bases, is directly reversed by the protein methyl guanine methyl transferase (MGMT), the bacterial equivalent of which is called ogt. This is an expensive process because each MGMT molecule can be used only once; that is, the reaction is stoichiometric rather than catalytic. A generalized response to methylating agents in bacteria is known as the adaptive response and confers a level of resistance to alkylating agents upon sustained exposure by upregulation of alkylation repair enzymes. The third type of DNA damage reversed by cells is certain methylation of the bases cytosine and adenine.

Single-strand Damage

When only one of the two strands of a double helix has a defect, the other strand can be used as a template to guide the correction of the damaged strand. In order to repair damage to one of the two paired molecules of DNA, there exist a number of excision repair mechanisms that remove the damaged nucleotide and replace it with an undamaged nucleotide complementary to that found in the undamaged DNA strand.

1. Base excision repair (BER), which repairs damage to a single base caused by oxidation, alkylation, hydrolysis, or deamination. The damaged base is removed by a DNA glycosylase. The "missing tooth" is then recognised by an enzyme called AP endonuclease, which cuts the Phosphodiester bond. The missing part is then resynthesized by a DNA polymerase, and a DNA ligase performs the final nick-sealing step.
2. Nucleotide excision repair (NER), which recognizes bulky, helix-distorting lesions such as pyrimidine dimers and 6,4 photoproducts. A specialized form of NER known as transcription-coupled repair deploys NER enzymes to genes that are being actively transcribed.
3. Mismatch repair (MMR), which corrects errors of DNA replication and recombination that result in mispaired (but undamaged) nucleotides.

Double-strand Breaks

Double-strand breaks, in which both strands in the double helix are severed, are particularly hazardous to the cell because they can lead to genome rearrangements. Three mechanisms exist to repair double-strand breaks (DSBs): non-homologous end joining (NHEJ), microhomology-mediated end joining (MMEJ), and homologous recombination. PVN Acharya noted that double-strand breaks and a "cross-linkage joining both strands at the same point is irreparable because neither strand can then serve as a template for repair. The cell will die in the next mitosis or in some rare instances, mutate." In NHEJ, DNA Ligase IV, a specialized DNA ligase that forms a complex with the cofactor XRCC4, directly joins the two ends. To guide accurate repair, NHEJ relies on short homologous sequences called microhomologies present on the single-stranded tails of the DNA ends to be joined. If these overhangs are compatible, repair is usually accurate. NHEJ can also introduce mutations during repair. Loss of damaged nucleotides at the break site can lead to deletions, and joining of nonmatching termini forms translocations. NHEJ is especially important before the cell has replicated its DNA, since there is no template available for repair by homologous recombination. There are "backup" NHEJ pathways in higher eukaryotes. Besides its role as a genome caretaker, NHEJ is required for joining hairpin-capped double-strand breaks induced during V(D)J recombination, the process that generates diversity in B-cell and T-cell receptors in the vertebrate immune system.

Homologous recombination requires the presence of an identical or nearly identical sequence to be used as a template for repair of the break. The enzymatic machinery responsible for this repair process is nearly identical to the machinery responsible for chromosomal crossover during meiosis. This pathway allows a damaged chromosome to be repaired using a sister chromatid (available in G2 after DNA replication) or a homologous chromosome as a template. DSBs caused by the replication machinery attempting to synthesize across a single-strand break or unrepaired lesion cause collapse of the replication fork and are typically repaired by recombination.

Topoisomerases introduce both single- and double-strand breaks in the course of changing the DNA's state of supercoiling, which is especially common in regions near an open replication fork. Such breaks are not considered DNA damage because they are a natural intermediate in the topoisomerase biochemical mechanism and are immediately repaired by the enzymes that created them.

A team of French researchers bombarded *Deinococcus radiodurans* to study the mechanism of double-strand break DNA repair in that organism. At least two copies of the genome, with random DNA breaks, can form DNA fragments through annealing. Partially overlapping fragments are then used for synthesis of homologous regions through a moving D-loop that can continue extension until they find complementary partner strands. In the final step there is crossover by means of RecA-dependent homologous recombination.

Translesion Synthesis

Translesion synthesis is a DNA damage tolerance process that allows the DNA replication machinery to replicate past DNA lesions such as thymine dimers or AP sites. It involves switching out regular DNA polymerases for specialized translesion polymerases (i.e. DNA polymerase IV or V, from the Y Polymerase family), often with larger active sites that can facilitate the insertion of bases opposite damaged nucleotides. The polymerase switching is thought to be mediated by, among other factors, the post-translational modification of the replication processivity factor PCNA.

Translesion synthesis polymerases often have low fidelity (high propensity to insert wrong bases) on undamaged templates relative to regular polymerases. However, many are extremely efficient at inserting correct bases opposite specific types of damage. For example, Pol ç mediates error-free bypass of lesions induced by UV irradiation, whereas Pol æ introduces mutations at these sites. From a cellular perspective, risking the introduction of point mutations during translesion synthesis may be preferable to resorting to more drastic mechanisms of DNA repair, which may cause gross chromosomal aberrations or cell death. In short, the process involves specialized polymerases either bypassing or repairing lesions at locations of stalled DNA replication. A bypass platform is provided to these polymerases by Proliferating cell nuclear antigen (PCNA). Under normal circumstances, PCNA bound to polymerases replicates the DNA. At a site of lesion, PCNA is ubiquitinated, or modified, by the RAD6/RAD18 proteins to provide a platform for the specialized polymerases to bypass the lesion and resume DNA replication.

Global Response to DNA Damage

Cells exposed to ionizing radiation, ultraviolet light or chemicals are prone to acquire multiple sites of bulky DNA lesions and double-strand breaks. Moreover, DNA damaging agents can damage other biomolecules such as proteins, carbohydrates, lipids, and RNA. The accumulation of damage, to be specific, double-strand breaks or adducts stalling the replication forks, are among known stimulation signals for a global response to DNA damage. The global response to damage is an act directed toward the cells' own preservation and triggers multiple pathways of macromolecular repair, lesion bypass, tolerance, or apoptosis. The common features of global response are induction of multiple genes, cell cycle arrest, and inhibition of cell division.

DNA Damage Checkpoints

After DNA damage, cell cycle checkpoints are activated. Checkpoint activation pauses the cell cycle and gives the cell time to repair the damage before continuing to divide. DNA damage checkpoints occur at the G1/S and G2/M boundaries. An intra-S checkpoint also exists. Checkpoint activation is controlled by two master kinases, ATM and ATR. ATM responds to DNA double-strand breaks and disruptions in chromatin structure, whereas ATR primarily responds to stalled replication forks. These kinases phosphorylate downstream targets in a signal transduction cascade, eventually leading to cell cycle arrest. A class of checkpoint mediator proteins including BRCA1, MDC1, and 53BP1 has also been identified. These proteins seem to be required for transmitting the checkpoint activation signal to downstream proteins. P53 is an important downstream target of ATM and ATR, as it is required for inducing apoptosis following DNA damage. At the G1/S checkpoint, p53 functions by deactivating the CDK2/cyclin E complex. Similarly, p21 mediates the G2/M checkpoint by deactivating the CDK1/cyclin B complex.

The Prokaryotic SOS Response

The SOS response is the term used to describe changes in gene expression in *Escherichia coli* and other bacteria in response to extensive DNA damage. The prokaryotic SOS system is regulated by two key proteins: LexA and RecA. The LexA homodimer is a transcriptional repressor that binds to operator sequences commonly referred to as SOS boxes.

In *Escherichia coli* it is known that LexA regulates transcription of approximately 48 genes including the lexA and recA genes. The SOS response is known to be widespread in the Bacteria domain, but it is mostly absent in some bacterial phyla, like the Spirochetes.

The most common cellular signals activating the SOS response are regions of single-stranded DNA (ssDNA), arising from stalled replication forks or double-strand breaks, which are processed by DNA helicase to separate the two DNA strands.

In the initiation step, RecA protein binds to ssDNA in an ATP hydrolysis driven reaction creating RecA–ssDNA filaments. RecA–ssDNA filaments activate LexA autoprotease activity, which ultimately leads to cleavage of LexA dimer and subsequent LexA degradation. The loss of LexA repressor induces transcription of the SOS genes and allows for further signal induction, inhibition of cell division and an increase in levels of proteins responsible for damage processing.

In *Escherichia coli*, SOS boxes are 20-nucleotide long sequences near promoters with palindromic structure and a high degree of sequence conservation. In other classes and phyla, the sequence of SOS boxes varies considerably, with different length and composition, but it is always highly conserved and one of the strongest short signals in the genome. The high information content of SOS boxes permits differential binding of LexA to different promoters and allows for timing of the SOS response. The lesion repair genes are induced at the beginning of SOS response. The error-prone translesion polymerases, for example, UmuCD'2 (also called DNA polymerase V), are induced later on as a last resort. Once the DNA damage is repaired or bypassed using polymerases or through recombination, the amount of single-stranded DNA in cells is decreased, lowering the amounts of RecA filaments decreases cleavage activity of LexA homodimer, which then binds to the SOS boxes near promoters and restores normal gene expression.

Eukaryotic Transcriptional Responses to DNA Damage

Eukaryotic cells exposed to DNA damaging agents also activate important defensive pathways by inducing multiple proteins involved in DNA repair, cell cycle checkpoint control, protein trafficking and degradation. Such genome wide transcriptional response is very complex and tightly regulated, thus allowing coordinated global response to damage.

Exposure of yeast *Saccharomyces cerevisiae* to DNA damaging agents results in overlapping but distinct transcriptional profiles. Similarities to environmental shock response indicates that a general global stress response pathway exist at the level of transcriptional activation. In contrast, different human cell types respond to damage differently indicating an absence of a common global response.

The probable explanation for this difference between yeast and human cells may be in the heterogeneity of mammalian cells. In an animal different types of cells are distributed among different organs that have evolved different sensitivities to DNA damage.

In general global response to DNA damage involves expression of multiple genes responsible for postreplication repair, homologous recombination, nucleotide excision repair, DNA damage checkpoint, global transcriptional activation, genes controlling mRNA decay, and many others.

A large amount of damage to a cell leaves it with an important decision: undergo apoptosis and die, or survive at the cost of living with a modified genome. An increase in tolerance to damage can lead to an increased rate of survival that will allow a greater accumulation of mutations. Yeast Rev1 and human polymerase ç are members of [Y family translesion DNA polymerases present during global response to DNA damage and are responsible for enhanced mutagenesis during a global response to DNA damage in eukaryotes.

DNA Repair and Evolution

The basic processes of DNA repair are highly conserved among both prokaryotes and eukaryotes and even among bacteriophage (viruses that infect bacteria); however, more complex organisms with more complex genomes have correspondingly more complex repair mechanisms.

The ability of a large number of protein structural motifs to catalyze relevant chemical reactions has played a significant role in the elaboration of repair mechanisms during evolution. For an extremely detailed review of hypotheses relating to the evolution of DNA repair.

The fossil record indicates that single-cell life began to proliferate on the planet at some point during the Precambrian period, although exactly when recognizably modern life first emerged is unclear. Nucleic acids became the sole and universal means of encoding genetic information, requiring DNA repair mechanisms that in their basic form have been inherited by all extant life forms from their common ancestor.

The emergence of Earth's oxygen-rich atmosphere (known as the "oxygen catastrophe") due to photosynthetic organisms, as well as the presence of potentially damaging free radicals in the cell due to oxidative phosphorylation, necessitated the evolution of DNA repair mechanisms that act specifically to counter the types of damage induced by oxidative stress.

Rate of Evolutionary Change

On some occasions, DNA damage is not repaired, or is repaired by an error-prone mechanism that results in a change from the original sequence. When this occurs, mutations may propagate into the genomes of the cell's progeny. Should such an event occur in a germ line cell that will eventually produce a gamete, the mutation has the potential to be passed on to the organism's offspring.

The rate of evolution in a particular species (or, in a particular gene) is a function of the rate of mutation. As a consequence, the rate and accuracy of DNA repair mechanisms have an influence over the process of evolutionary change.

Senescence

Senescence or biological aging is the change in the biology of an organism as it ages after its maturity. Such changes range from those affecting its cells and their function to that of the whole organism. There are a number of theories as to why senescence occurs, including ones that claim it is programmed by gene expression changes and that it is the accumulative damage of biological processes.

The word *senescence* is derived from the Latin word *senex*, meaning *old man*, *old age*, or *advanced in age*.

Cellular Senescence

Cellular senescence is the phenomenon by which normal diploid cells lose the ability to divide, normally after about 50 cell divisions in vitro. Some cells become senescent after fewer replications cycles as a result of DNA double strand breaks, toxins, etc.

This phenomenon is also known as "replicative senescence", the "Hayflick phenomenon", or the Hayflick limit in honour of Dr. Leonard Hayflick who was the first to publish this information in 1965. In response to DNA damage (including shortened telomeres), cells either age or self-destruct (apoptosis, programmed cell death) if the damage cannot be repaired.

In this 'cellular suicide', the death of one cell, or more, may benefit the organism as a whole. For example, in plants the death of the water-conducting xylem cells (tracheids and vessel elements) allows the cells to function more efficiently and so deliver water to the upper parts of a plant.

Aging of the Whole Organism

Organismal senescence is the aging of whole organisms. In general, aging is characterized by the declining ability to respond to stress, increased homeostatic imbalance, and increased risk of aging-associated diseases. Death is the ultimate consequence of aging, though "old age" is not a scientifically recognized cause of death because there is always a specific proximal cause, such as cancer, heart disease, or liver failure. Aging of whole organisms is therefore a complex process that can be defined as "a progressive deterioration of physiological function, an intrinsic age-related process of loss of viability and increase in vulnerability".

Differences in maximum life span among species correspond to different "rates of aging". For example, inherited differences in the rate of aging make a mouse elderly at 3 years and a human elderly at 80 years. These genetic differences affect a variety of physiological processes, including the efficiency of DNA repair, antioxidant enzymes, and rates of free radical production.

Senescence of the organism gives rise to the Gompertz–Makeham law of mortality, which says that mortality rate rises rapidly with age.

Some animals, such as some reptiles and fish, age slowly (negligible senescence) and exhibit very long lifespans. Some even exhibit "negative senescence", in which mortality falls with age, in disagreement with the Gompertz–Makeham "law".

Whether replicative senescence (Hayflick limit) plays a causative role in organismal aging is at present an active area of investigation.

Theories of Aging

The process of senescence is complex, and may derive from a variety of different mechanisms and exist for a variety of different reasons. However, senescence is not universal, and scientific evidence suggests that cellular senescence evolved in certain species because it prevents the onset of cancer. In a few simple species, such as Hydra, senescence is negligible and cannot be detected. All such species have no "post-mitotic" cells; they reduce the effect of damaging free radicals by cell division and dilution. Such species are not immortal, however, as they will eventually fall prey to trauma or disease. Moreover, average lifespans can vary greatly within and between species. This suggests that both genetic and environmental factors contribute to aging. In general, theories that explain senescence have been divided between the programmed and stochastic theories of aging. Programmed theories imply that aging is regulated by biological clocks operating throughout the lifespan. This regulation would depend on changes in gene expression that affect the systems responsible for maintenance, repair, and defense responses. Stochastic theories blame environmental impacts on living organisms that induce cumulative damage at various levels as the cause of aging, examples of which ranging from damage to DNA, damage to tissues and cells by oxygen radicals (widely known as free radicals countered by the even more well-known antioxidants), and cross-linking.

However, aging is seen as a progressive failure of homeodynamics (homeostasis) involving genes for the maintenance and repair, stochastic events leading to molecular damage and molecular heterogeneity, and chance events determining the probability of death. Since complex and interacting systems of maintenance and repair comprise the homeodynamic (old term: homeostasis) space of a biological system, aging is considered to be a progressive shrinkage of homeodynamic space mainly due to increased molecular heterogeneity.

Evolutionary Theories

A gene can be expressed at various life-stages. Therefore, natural selection can support lethal and harmful alleles, if their expression occurs after reproduction. Senescence may be the product of such selection. In addition, aging is believed to have evolved because of the increasingly smaller probability of an organism still being alive at older age, due to predation and accidents, both of which may be random and age-invariant.

It is thought that strategies that result in a higher reproductive rate at a young age, but shorter overall lifespan, result in a higher lifetime reproductive success and are therefore favoured by natural selection. In essence, aging is, therefore, the result of investing resources in reproduction, rather than maintenance of the body (the “Disposable Soma” theory), in light of the fact that accidents, predation, and disease will eventually kill the organism no matter how much energy is devoted to repair of the body. Various other theories of aging exist, and are not necessarily mutually exclusive.

The geneticist J. B. S. Haldane wondered why the dominant mutation that causes Huntington’s disease remained in the population, and why natural selection had not eliminated it. The onset of this neurological disease is (on average) at age 45 and is invariably fatal within 10–20 years. Haldane assumed that, in human prehistory, few survived until age 45. Since few were alive at older ages and their contribution to the next generation was therefore small relative to the large cohorts of younger age groups, the force of selection against such late-acting deleterious mutations was correspondingly small. However, if a mutation affected younger individuals, selection against it would be strong. Therefore, late-acting deleterious mutations could accumulate in populations over evolutionary time through genetic drift. This principle has been demonstrated experimentally. And it is these later-acting deleterious mutations that are believed to cause—even allow—age-related mortality.

Peter Medawar formalised this observation in his mutation accumulation theory of aging. "The force of natural selection weakens with increasing age—even in a theoretically immortal population, provided only that it is exposed to real hazards of mortality. If a genetic disaster... happens late enough in individual life, its consequences may be completely unimportant". The 'real hazards of mortality' are, in typical circumstances, predation, disease, and accidents. So, even an immortal population, whose fertility does not decline with time, will have fewer individuals alive in older age groups. This is called 'extrinsic mortality'. Young cohorts, not depleted in numbers yet by extrinsic mortality, contribute far more to the next generation than the few remaining older cohorts, so the force of selection against late-acting deleterious mutations, which affect only these few older individuals, is very weak. The mutations may not be selected against, therefore, and may spread over evolutionary time into the population.

The major testable prediction made by this model is that species that have high extrinsic mortality in nature will age more quickly and have shorter intrinsic lifespans. This is borne out among mammals, the best-studied in terms of life history. There is a correlation among mammals between body size and lifespan, such that larger species live longer than smaller species in controlled/optimum conditions, but there are notable exceptions.

For instance, many bats and rodents are of similar size, yet bats live much longer. For instance, the little brown bat, half the size of a mouse, can live 30 years in the wild. A mouse will live 2–3 years even with optimum conditions. The explanation is that bats have fewer predators, so therefore low extrinsic mortality. Thus, more individuals survive to later ages, so the force of selection against late-acting deleterious mutations is stronger. Fewer late-acting deleterious mutations = slower aging = longer lifespan. Birds are also warm-blooded and are of similar size to many small mammals, yet live often 5–10 times as long. They have fewer predation pressures compared with ground-dwelling mammals. And seabirds, which, in general, have the fewest predators of all birds, live longest.

Also, when examining the body-size vs. lifespan relationship, one will observe that predator mammals tend to have longer lifespans than prey animals in a controlled environment such as a zoo or nature reserve. The explanation for the long lifespans of primates (such as humans, monkeys, and apes) relative to body size is that their intelligence and often sociality helps them avoid becoming prey. Being a predator, being smart, and working together all reduce extrinsic mortality. Another evolutionary theory of aging was proposed by George C. Williams and involves antagonistic pleiotropy. A single gene may affect multiple traits. Some traits that increase fitness early in life may also have negative effects later in life.

But, because many more individuals are alive at young ages than at old ages, even small positive effects early can be strongly selected for, and large negative effects later may be very weakly selected against. Williams suggested the following example: Perhaps a gene codes for calcium deposition in bones, which promotes juvenile survival and will therefore be favored by natural selection; however, this same gene promotes calcium deposition in the arteries, causing negative effects in old age. Therefore, negative effects in old age may reflect the result of natural selection for pleiotropic genes that are beneficial early in life. In this case, fitness is relatively high when Fisher's reproductive value is high and relatively low when Fisher's reproductive value is low.

Gene Regulation

A number of genetic components of aging have been identified using model organisms, ranging from the simple budding yeast *Saccharomyces cerevisiae* to worms such as *Caenorhabditis elegans* and fruit flies (*Drosophila melanogaster*). Study of these organisms has revealed the presence of at least two conserved aging pathways.

One of these pathways involves the gene *Sir2*, a NAD+-dependent histone deacetylase. In yeast, Sir2 is required for genomic silencing at three loci: The yeast mating loci, the telomeres and the ribosomal DNA (rDNA). In some species of yeast, replicative aging may be partially caused by homologous recombination between rDNA repeats; excision of rDNA repeats results in the formation of extrachromosomal rDNA circles (ERCs). These ERCs replicate and preferentially segregate to the mother cell during cell division, and are believed to result in cellular senescence by titrating away (competing for) essential nuclear factors. ERCs have not been observed in other species (nor even all strains of the same yeast species) of yeast (which also display replicative senescence), and ERCs are not believed to contribute to aging in higher organisms such as humans (they have not been shown to accumulate in mammals in a similar manner to yeast). Extrachromosomal circular DNA (eccDNA) has been found in worms, flies, and humans. The origin and role of eccDNA in aging, if any, is unknown.

Despite the lack of a connection between circular DNA and aging in higher organisms, extra copies of Sir2 are capable of extending the lifespan of both worms and flies (though, in flies, this finding has not been replicated by other investigators, and the activator of Sir2 resveratrol does not reproducibly increase lifespan in either species). Whether the Sir2 homologues in higher organisms have any role in lifespan is unclear, but the human SIRT1 protein has been demonstrated to deacetylate p53, Ku70, and the forkhead family of transcription factors. SIRT1 can also regulate acetylates such as CBP/p300, and has been shown to deacetylate specific histone residues.

RAS1 and RAS2 also affect aging in yeast and have a human homologue. RAS2 overexpression has been shown to extend lifespan in yeast.

Other genes regulate aging in yeast by increasing the resistance to oxidative stress. Superoxide dismutase, a protein that protects against the effects of mitochondrial free radicals, can extend yeast lifespan in stationary phase when overexpressed.

In higher organisms, aging is likely to be regulated in part through the insulin/IGF-1 pathway. Mutations that affect insulin-like signaling in worms, flies, and the growth hormone/IGF1 axis in mice are associated with extended lifespan. In yeast, Sir2 activity is regulated by the nicotinamidase PNC1. PNC1 is transcriptionally upregulated under stressful conditions such as caloric restriction, heat shock, and osmotic shock. By converting nicotinamide to niacin, nicotinamide is removed, inhibiting the activity of Sir2. A nicotinamidase found in humans, known as PBEF, may serve a similar function, and a secreted form of PBEF known as visfatin may help to regulate serum insulin levels. It is not known, however, whether these mechanisms also exist in humans, since there are obvious differences in biology between humans and model organisms.

Sir2 activity has been shown to increase under calorie restriction. Due to the lack of available glucose in the cells, more NAD+ is available and can activate Sir2. Resveratrol, a stilbenoid found in the skin of red grapes, was reported to extend the lifespan of yeast, worms, and flies (the lifespan extension in flies and worms have proved irreproducible by independent investigators). It has been shown to activate Sir2 and therefore mimics the effects of calorie restriction, if one accepts that caloric restriction is indeed dependent on Sir2.

Gene expression is imperfectly controlled, and it is possible that random fluctuations in the expression levels of many genes contribute to the aging process as suggested by a study of such genes in yeast. Individual cells, which are genetically identical, none-the-less can have substantially different responses to outside stimuli, and markedly different lifespans, indicating the epigenetic factors play an important role in gene expression and aging as well as genetic factors. According to the GenAge database of aging-related genes there are over 700 genes associated with aging in model organisms: 555 in the soil roundworm (*Caenorhabditis elegans*), 87 in the bakers' yeast (*Saccharomyces cerevisiae*), 75 in the fruit fly (*Drosophila melanogaster*) and 68 in the mouse (*Mus musculus*).

Cellular Senescence

As noted above, senescence is not universal, and senescence is not observed in single-celled organisms that reproduce through the process of cellular mitosis. Moreover, cellular senescence is not observed in several organisms, including perennial plants, sponges, corals, and lobsters. In those species where cellular senescence is observed, cells eventually become post-mitotic when they can no longer replicate themselves through the process of cellular mitosis; i.e., cells experience *replicative senescence*. How and why some cells become post-mitotic in some species has been the subject of much research and speculation, but (as noted above) it is widely believed that cellular senescence evolved as a way to prevent the onset and spread of cancer. Somatic cells that have divided many times will have accumulated DNA mutations and would therefore be in danger of becoming cancerous if cell division continued.

Lately, the role of telomeres in cellular senescence has aroused general interest, especially with a view to the possible genetically adverse effects of cloning. The successive shortening of the chromosomal telomeres with each cell cycle is also believed to limit the number of divisions of the cell, thus contributing to aging. There have, on the other hand, also been reports that cloning could alter the shortening of telomeres. Some cells do not age and are, therefore, described as being "biologically immortal". It is theorized by some that when it is discovered exactly what allows these cells, whether it be the result of telomere lengthening or not, to divide without limit that it will be possible to genetically alter other cells to have the same capability. It is further theorized that it will eventually be possible to genetically engineer all cells in the human body to have this capability by employing gene therapy and, therefore, stop or reverse aging, effectively making the entire organism potentially immortal.

Cancer cells are usually immortal. In about 85% of tumors, this evasion of cellular senescence is the result of up-activation of their telomerase genes. This simple observation suggests that reactivation of telomerase in healthy individuals could greatly increase their cancer risk. Whether cell senescence plays any role in organismal aging is at present unknown, and is an active area of investigation. Mouse mutants lacking telomerase do not immediately show accelerated aging.

Chemical Damage

One of the earliest aging theories was the *Rate of Living Hypothesis* described by Raymond Pearl in 1928(based on earlier work by Max Rubner), which states that fast basal metabolic rate corresponds to short maximum life span. While there may be some validity to the idea that for various types of specific damage detailed below that are by-products of metabolism, all other things being equal, a fast metabolism may reduce lifespan, in general this theory does not adequately explain the differences in lifespan either within, or between, species. Calorically-restricted animals process as much, or more, calories per gram of body mass, as their *ad libitum* fed counterparts, yet exhibit substantially longer lifespans.. Similarly, metabolic rate is a poor predictor of lifespan for birds, bats and other species that, it is presumed, have reduced mortality from predation, and therefore have evolved long lifespans even in the presence of very high metabolic rates. More recently, it was shown that, when modern statistical methods for correcting for the effects of body size and phylogeny are employed, metabolic rate does not correlate with longevity in mammals or birds.

With respect to specific types of chemical damage caused by metabolism, it is suggested that damage to long-lived biopolymers, such as structural proteins or DNA, caused by ubiquitous chemical agents in the body such as oxygen and sugars, are in part responsible for aging. The damage can include breakage of biopolymer chains, cross-linking of biopolymers, or chemical attachment of unnatural substituents (haptens) to biopolymers.

Under normal aerobic conditions, approximately 4% of the oxygen metabolized by mitochondria is converted to superoxide ion, which can subsequently be converted to hydrogen peroxide, hydroxyl radical and eventually other reactive species including other peroxides and singlet oxygen, which can, in turn, generate free radicals capable of damaging structural proteins and DNA. Certain metal ions found in the body, such as copper and iron, may participate in the process. (In Wilson's disease, a hereditary defect that causes the body to retain copper, some of the symptoms resemble accelerated senescence.) These processes are termed *oxidative damage* and are linked to the benefits of nutritionally derived polyphenol antioxidants.

Sugars such as glucose and fructose can react with certain amino acids such as lysine and arginine and certain DNA bases such as guanine to produce sugar adducts, in a process called *glycation*. These adducts can further rearrange to form reactive species, which can then cross-link the structural proteins or DNA to similar biopolymers or other biomolecules such as non-structural proteins. People with diabetes, who have elevated blood sugar, develop senescence-associated disorders much earlier than the general population, but can delay such disorders by rigorous control of their blood sugar levels. There is evidence that sugar damage is linked to oxidant damage in a process termed *glycoxidation*.

Free radicals can damage proteins, lipids or DNA. Glycation mainly damages proteins. Damaged proteins and lipids accumulate in lysosomes as lipofuscin. Chemical damage to structural proteins can lead to loss of function; for example, damage to collagen of blood vessel walls can lead to vessel-wall stiffness and, thus, hypertension, and vessel wall thickening and reactive tissue formation (atherosclerosis); similar processes in the kidney can lead to renal failure. Damage to enzymes reduces cellular functionality. Lipid peroxidation of the inner mitochondrial membrane reduces the electric potential and the ability to generate energy. It is probably no accident that nearly all of the so-called "accelerated aging diseases" are due to defective DNA repair enzymes.

It is believed that the impact of alcohol on aging can be partly explained by alcohol's activation of the HPA axis, which stimulates glucocorticoid secretion, long-term exposure to which produces symptoms of aging.

Reliability Theory

Reliability theory suggests that biological systems start their adult life with a high load of initial damage. Reliability theory is a general theory about systems failure. It allows researchers to predict the age-related failure kinetics for a system of given architecture (reliability structure) and given reliability of its components. Reliability theory predicts that even those systems that composed entirely of non-aging elements (with a constant failure rate) will nevertheless deteriorate (fail more often) with age, if these systems are redundant in irreplaceable elements. Aging, therefore, is a direct consequence of systems.

Reliability theory also predicts the late-life mortality deceleration with subsequent leveling-off, as well as the late-life mortality plateaus, as an inevitable consequence of redundancy exhaustion at extreme old ages.

The theory explains why mortality rates increase exponentially with age (the Gompertz law) in many species, by taking into account the initial flaws (defects) in newly formed systems. It also explains why organisms "prefer" to die according to the Gompertz law, while technical devices usually fail according to the Weibull (power) law. Reliability theory allows to specify conditions when organisms die according to the Weibull distribution: Organisms should be relatively free of initial flaws and defects.

The theory makes it possible to find a general failure law applicable to all adult and extreme old ages, where the Gompertz and the Weibull laws are just special cases of this more general failure law. The theory explains why relative differences in mortality rates of compared populations (within a given species) vanish with age (compensation law of mortality), and mortality convergence is observed due to the exhaustion of initial differences in redundancy levels.

Miscellaneous

Recently, a kind of early senescence has been alleged to be a possible unintended outcome of early cloning experiments. The issue was raised in the case of Dolly the sheep, following her death from a contagious lung disease. The claim that Dolly's early death involved premature senescence has been vigorously contested, and Dolly's creator, Dr. Ian Wilmut has expressed the view that her illness and death were probably unrelated to the fact that she was a clone. A set of rare hereditary (genetic) disorders, each called progeria, has been known for some time. Sufferers exhibit symptoms resembling accelerated aging, including wrinkled skin. The cause of Hutchinson–Gilford progeria syndrome was reported in the journal *Nature* in May 2003. This report suggests that DNA damage, not oxidative stress, is the cause of this form of accelerated aging.

Transcription

Transcription is the process of creating a complementary RNA copy of a sequence of DNA. Both RNA and DNA are nucleic acids, which use base pairs of nucleotides as a complementary language that can be converted back and forth from DNA to RNA by the action of the correct enzymes. During transcription, a DNA sequence is read by RNA polymerase, which produces a complementary, antiparallel RNA strand. As opposed to DNA replication, transcription results in an RNA complement that includes uracil (U) in all instances where thymine (T) would have occurred in a DNA complement.

Transcription can be explained easily in 4 or 5 simple steps, each moving like a wave along the DNA.

1. RNA polymerase unwinds/"unzips" the DNA by breaking the hydrogen bonds between complimentary nucleotides.
2. RNA nucleotides are paired with complementary DNA bases.
3. RNA sugar-phosphate backbone forms with assistance from RNA polymerase.
4. Hydrogen bonds of the untwisted RNA+DNA helix break, freeing the newly synthesized RNA strand.
5. If the cell has a nucleus, the RNA is further processed and then moves through the small nuclear pores to the cytoplasm.

Transcription is the first step leading to gene expression. The stretch of DNA transcribed into an RNA molecule is called a *transcription unit* and encodes at least one gene. If the gene transcribed encodes a protein, the result of transcription is messenger RNA (mRNA), which will then be used to create that protein via the process of translation. Alternatively, the transcribed gene may encode for either ribosomal RNA (rRNA) or transfer RNA (tRNA), other components of the protein-assembly process, or other ribozymes. A DNA transcription unit encoding for a protein contains not only the sequence that will eventually be directly translated into the protein (the *coding sequence*) but also *regulatory sequences* that direct and regulate the synthesis of that protein. The regulatory sequence before (upstream from) the coding sequence is called the five prime untranslated region (5'UTR), and the sequence following (downstream from) the coding sequence is called the three prime untranslated region (3'UTR).

Transcription has some proofreading mechanisms, but they are fewer and less effective than the controls for copying DNA; therefore, transcription has a lower copying fidelity than DNA replication.

As in DNA replication, DNA is read from 3' → 5' during transcription. Meanwhile, the complementary RNA is created from the 5' → 3' direction. This means its 5' end is created first in base pairing. Although DNA is arranged as two antiparallel strands in a double helix, only one of the two DNA strands, called the template strand, is used for transcription. This is because RNA is only single-stranded, as opposed to double-stranded DNA. The other DNA strand is called the coding strand, because its sequence is the same as the newly created RNA transcript (except for the substitution of uracil for thymine). The use of only the 3' → 5' strand eliminates the need for the Okazaki fragments seen in DNA replication.

Transcription is divided into 5 stages: *pre-initiation*, *initiation*, *promoter clearance*, *elongation* and *termination*.

Major Steps

Pre-initiation

In eukaryotes, RNA polymerase, and therefore the initiation of transcription, requires the presence of a core promoter sequence in the DNA. Promoters are regions of DNA that promote transcription and, in eukaryotes, are found at -30, -75, and -90 base pairs upstream from the start site of transcription. Core promoters are sequences within the promoter that are essential for transcription initiation. RNA polymerase is able to bind to core promoters in the presence of various specific transcription factors.

The most common type of core promoter in eukaryotes is a short DNA sequence known as a TATA box, found 25-30 base pairs upstream from the start site of transcription. The TATA box, as a core promoter, is the binding site for a transcription factor known as TATA-binding protein (TBP), which is itself a subunit of another transcription factor, called Transcription Factor II D (TFIID). After TFIID binds to the TATA box via the TBP, five more transcription factors and RNA polymerase combine around the TATA box in a series of stages to form a preinitiation complex. One transcription factor, DNA helicase, has helicase activity and so is involved in the separating of opposing strands of double-stranded DNA to provide access to a single-stranded

DNA template. However, only a low, or basal, rate of transcription is driven by the preinitiation complex alone. Other proteins known as activators and repressors, along with any associated coactivators or corepressors, are responsible for modulating transcription rate.

Thus, preinitiation complex contains:

1. Core Promoter Sequence
2. Transcription Factors
3. DNA Helicase
4. RNA Polymerase
5. Activators and Repressors.

The transcription preinitiation in archaea is, in essence, homologous to that of eukaryotes, but is much less complex. The archaeal preinitiation complex assembles at a TATA-box binding site; however, in archaea, this complex is composed of only RNA polymerase II, TBP, and TFB (the archaeal homologue of eukaryotic transcription factor II B (TFIIB)).

Initiation

In bacteria, transcription begins with the binding of RNA polymerase to the promoter in DNA. RNA polymerase is a core enzyme consisting of five subunits: 2 α subunits, 1 β subunit, 1 β' subunit, and 1 ω subunit. At the start of initiation, the core enzyme is associated with a sigma factor that aids in finding the appropriate -35 and -10 base pairs downstream of promoter sequences.

Transcription initiation is more complex in eukaryotes. Eukaryotic RNA polymerase does not directly recognize the core promoter sequences. Instead, a collection of proteins called transcription factors mediate the binding of RNA polymerase and the initiation of transcription. Only after certain transcription factors are attached to the promoter does the RNA polymerase bind to it. The completed assembly of transcription factors and RNA polymerase bind to the promoter, forming a transcription initiation complex. Transcription in the archaea domain is similar to transcription in eukaryotes.

Promoter Clearance

After the first bond is synthesized, the RNA polymerase must clear the promoter. During this time there is a tendency to release the RNA transcript and produce truncated transcripts. This is called *abortive initiation* and is common for both eukaryotes and prokaryotes. Abortive initiation continues to occur until the σ factor rearranges, resulting in the transcription elongation complex (which gives a 35 bp moving footprint). The σ factor is released before 80 nucleotides of mRNA are synthesized. Once the transcript reaches approximately 23 nucleotides, it no longer slips and elongation can occur. This, like most of the remainder of transcription, is an energy-dependent process, consuming adenosine triphosphate (ATP). Promoter clearance coincides with phosphorylation of serine 5 on the carboxy terminal domain of RNA Pol in eukaryotes, which is phosphorylated by TFIIH.

Elongation

One strand of the DNA, the *template strand* (or noncoding strand), is used as a template for RNA synthesis. As transcription proceeds, RNA polymerase traverses the template strand and uses base pairing complementarity with the DNA template to create an RNA copy. Although RNA polymerase traverses the template strand from 3' → 5', the coding (non-template) strand and newly-formed RNA can also be used as reference points, so transcription can be described as occurring 5' → 3'.

This produces an RNA molecule from 5' → 3', an exact copy of the coding strand (except that thymines are replaced with uracils, and the nucleotides are composed of a ribose (5-carbon) sugar where DNA has deoxyribose (one less oxygen atom) in its sugar-phosphate backbone).

Unlike DNA replication, mRNA transcription can involve multiple RNA polymerases on a single DNA template and multiple rounds of transcription (amplification of particular mRNA), so many mRNA molecules can be rapidly produced from a single copy of a gene.

Elongation also involves a proofreading mechanism that can replace incorrectly incorporated bases. In eukaryotes, this may correspond with short pauses during transcription that allow appropriate RNA editing factors to bind. These pauses may be intrinsic to the RNA polymerase or due to chromatin structure.

Termination

Bacteria use two different strategies for transcription termination. In Rho-independent transcription termination, RNA transcription stops when the newly synthesized RNA molecule forms a G-C-rich hairpin loop followed by a run of Us. When the hairpin forms, the mechanical stress breaks the weak rU-dA bonds, now filling the DNA-RNA hybrid. This pulls the poly-U transcript out of the active site of the RNA polymerase, in effect, terminating transcription. In the "Rho-dependent" type of termination, a protein factor called "Rho" destabilizes the interaction between the template and the mRNA, thus releasing the newly synthesized mRNA from the elongation complex.

Transcription termination in eukaryotes is less understood but involves cleavage of the new transcript followed by template-independent addition of As at its new 3' end, in a process called polyadenylation.

Measuring and Detecting Transcription

Transcription can be measured and detected in a variety of ways:

- Nuclear Run-on assay: measures the relative abundance of newly formed transcripts
- RNase protection assay and ChIP-Chip of RNAP: detect active transcription sites
- RT-PCR: measures the absolute abundance of total or nuclear RNA levels, which may however differ from transcription rates
- DNA microarrays: measures the relative abundance of the global total or nuclear RNA levels; however, these may differ from transcription rates
- In situ hybridization: detects the presence of a transcript
- MS2 tagging: by incorporating RNA stem loops, such as MS2, into a gene, these become incorporated into newly synthesized RNA. The stem loops can then be detected using a fusion of GFP and the MS2 coat protein, which has a high affinity, sequence-specific interaction with the MS2 stem loops. The recruitment of GFP to the site of transcription is visualised as a single fluorescent spot. This remarkable new approach has revealed that transcription occurs in discontinuous bursts, or pulses. With the notable exception of in situ techniques, most other methods provide cell population averages, and are not capable of detecting this fundamental property of genes.
- Northern blot: the traditional method, and until the advent of RNA-Seq, the most quantitative
- RNA-Seq: applies next-generation sequencing techniques to sequence whole transcriptomes, which allows the measurement of relative abundance of RNA, as well as the detection of additional variations such as fusion genes, post-translational edits and novel splice sites.

Transcription Factories

Active transcription units are clustered in the nucleus, in discrete sites called transcription factories or euchromatin. Such sites can be visualized by allowing engaged polymerases to extend their transcripts in tagged precursors (Br-UTP or Br-U) and immuno-labelling the tagged nascent RNA. Transcription factories can also be localized using fluorescence in situ hybridization or marked by antibodies directed against polymerases. There are ~10,000 factories in the nucleoplasm of a HeLa cell, among which are ~8,000 polymerase II factories and ~2,000 polymerase III factories. Each polymerase II factory contains ~8 polymerases. As most active transcription units are associated with only one polymerase, each factory usually contains ~8 different transcription units. These units might be associated through promoters and/or enhancers, with loops forming a 'cloud' around the factor.

History

A molecule that allows the genetic material to be realized as a protein was first hypothesized by François Jacob and Jacques Monod. RNA synthesis by RNA polymerase was established *in vitro* by several laboratories by 1965; however, the RNA synthesized by these enzymes had properties that suggested the existence of an additional factor needed to terminate transcription correctly.

In 1972, Walter Fiers became the first person to actually prove the existence of the terminating enzyme.

Roger D. Kornberg won the 2006 Nobel Prize in Chemistry "for his studies of the molecular basis of eukaryotic transcription".

Reverse Transcription

Some viruses (such as HIV, the cause of AIDS), have the ability to transcribe RNA into DNA. HIV has an RNA genome that is duplicated into DNA. The resulting DNA can be merged with the DNA genome of the host cell. The main enzyme responsible for synthesis of DNA from an RNA template is called reverse transcriptase. In the case of HIV, reverse transcriptase is responsible for synthesizing a complementary DNA strand (cDNA) to the viral RNA genome. An associated enzyme, ribonuclease H, digests the RNA strand, and reverse transcriptase synthesises a complementary strand of DNA to form a double helix DNA structure.

This cDNA is integrated into the host cell's genome via another enzyme (integrase) causing the host cell to generate viral proteins that reassemble into new viral particles. In HIV, subsequent to this, the host cell undergoes programmed cell death, apoptosis of T cells. However, in other retroviruses, the host cell remains intact as the virus buds out of the cell.

Some eukaryotic cells contain an enzyme with reverse transcription activity called telomerase. Telomerase is a reverse transcriptase that lengthens the ends of linear chromosomes. Telomerase carries an RNA template from which it synthesizes DNA repeating sequence, or "junk" DNA. This repeated sequence of DNA is important because, every time a linear chromosome is duplicated, it is shortened in length. With "junk" DNA at the ends of chromosomes, the shortening eliminates some of the non-essential, repeated sequence rather than the protein-encoding DNA sequence farther away from the chromosome end. Telomerase is often activated in cancer cells to enable cancer cells to duplicate their genomes indefinitely without losing important protein-coding DNA sequence. Activation of telomerase could be part of the process that allows cancer cells to become *immortal*. However, the true *in vivo* significance of telomerase has still not been empirically proven.

Ribonucleic Acid

Ribonucleic acid (RNA) is one of the three major macromolecules (along with DNA and proteins) that are essential for all known forms of life. Like DNA, RNA is made up of a long chain of components called nucleotides. Each nucleotide consists of a nucleobase (sometimes called a nitrogenous base), a ribose sugar, and a phosphate group. The sequence of nucleotides allows RNA to encode genetic information. For example, some viruses use RNA instead of DNA as their genetic material, and all organisms use messenger RNA (mRNA) to carry the genetic information that directs the synthesis of proteins.

Like proteins, some RNA molecules play an active role in cells by catalyzing biological reactions, controlling gene expression, or sensing and communicating responses to cellular signals. One of these active processes is protein synthesis, a universal function whereby mRNA molecules direct the assembly of proteins on ribosomes. This process uses transfer RNA (tRNA) molecules to deliver amino acids to the ribosome, where ribosomal RNA (rRNA) links amino acids together to form proteins.

The chemical structure of RNA is very similar to that of DNA, with two differences—(a) RNA contains the sugar ribose while DNA contains the slightly different sugar deoxyribose (a type of ribose that lacks one oxygen atom), and (b) RNA has the nucleobase uracil while DNA contains thymine (uracil and thymine have similar base-pairing properties). Unlike DNA, most RNA molecules are single-stranded. Single-stranded RNA molecules adopt very complex three-dimensional structures, since they are not restricted to the repetitive double-helical form of double-stranded DNA. RNA is made within living cells by RNA polymerases, enzymes that act to copy a DNA or RNA template into a new RNA strand through processes known as transcription or RNA replication, respectively.

Comparison with DNA

RNA and DNA are both nucleic acids, but differ in three main ways. First, unlike DNA, which is, in general, double-stranded, RNA is a single-stranded molecule in many of its biological roles and has a much shorter chain of nucleotides. Second, while DNA contains *deoxyribose*, RNA contains *ribose* (in deoxyribose there is no hydroxyl group attached to the pentose ring in the 2' position). These hydroxyl groups make RNA less stable than DNA because it is more prone to hydrolysis. Third, the complementary base to adenine is not thymine, as it is in DNA, but rather uracil, which is an unmethylated form of thymine.

Like DNA, most biologically active RNAs, including mRNA, tRNA, rRNA, snRNAs, and other non-coding RNAs, contain self-complementary sequences that allow parts of the RNA to fold and pair with itself to form double helices. Structural analysis of these RNAs has revealed that they are highly structured. Unlike DNA, their structures do not consist of long double helices but rather collections of short helices packed together into structures akin to proteins. In this fashion, RNAs can achieve chemical catalysis, like enzymes. For instance, determination of the structure of the ribosome—an enzyme that catalyzes peptide bond formation—revealed that its active site is composed entirely of RNA.

Structure

Each nucleotide in RNA contains a ribose sugar, with carbons numbered 1' through 5'. A base is attached to the 1' position, in general, adenine (A), cytosine (C), guanine (G), or uracil (U). Adenine and guanine are purines, cytosine, and uracil are pyrimidines. A phosphate group is attached to the 3' position of one ribose and the 5' position of the next. The phosphate groups have a negative charge each at physiological pH, making RNA a charged molecule (polyanion). The bases may form hydrogen bonds between cytosine and guanine, between adenine and uracil and between guanine and uracil. However, other interactions are possible, such as a group of adenine bases binding to each other in a bulge, or the GNRA tetraloop that has a guanine–adenine base-pair.

An important structural feature of RNA that distinguishes it from DNA is the presence of a hydroxyl group at the 2' position of the ribose sugar. The presence of this functional group causes the helix to adopt the A-form geometry rather than the B-form most commonly observed in DNA. This results in a very deep and narrow major groove and a shallow and wide minor groove.

A second consequence of the presence of the 2'-hydroxyl group is that in conformationally flexible regions of an RNA molecule (that is, not involved in formation of a double helix), it can chemically attack the adjacent phosphodiester bond to cleave the backbone. RNA is transcribed with only four bases (adenine, cytosine, guanine and uracil), but these bases and attached sugars can be modified in numerous ways as the RNAs mature.

Pseudouridine (Ø), in which the linkage between uracil and ribose is changed from a C–N bond to a C–C bond, and ribothymidine (T) are found in various places (the most notable ones being in the TØC loop of tRNA). Another notable modified base is hypoxanthine, a deaminated adenine base whose nucleoside is called inosine (I). Inosine plays a key role in the wobble hypothesis of the genetic code. There are nearly 100 other naturally occurring modified nucleosides, of which pseudouridine and nucleosides with 2'-Omethylribose are the most common. The specific roles of many of these modifications in RNA are not fully understood. However, it is notable that, in ribosomal RNA, many of the post-transcriptional modifications occur in highly functional regions, such as the peptidyl transferase center and the subunit interface, implying that they are important for normal function.

The functional form of single stranded RNA molecules, just like proteins, frequently requires a specific tertiary structure.

The scaffold for this structure is provided by secondary structural elements that are hydrogen bonds within the molecule. This leads to several recognizable "domains" of secondary structure like hairpin loops, bulges, and internal loops. Since RNA is charged, metal ions such as Mg^{2+} are needed to stabilise many secondary and tertiary structures.

Synthesis

Synthesis of RNA is usually catalyzed by an enzyme—RNA polymerase—using DNA as a template, a process known as transcription. Initiation of transcription begins with the binding of the enzyme to a promoter sequence in the DNA (usually found "upstream" of a gene). The DNA double helix is unwound by the helicase activity of the enzyme. The enzyme then progresses along the template strand in the 3' to 5' direction, synthesizing a complementary RNA molecule with elongation occurring in the 5' to 3' direction. The DNA sequence also dictates where termination of RNA synthesis will occur.

RNAs are often modified by enzymes after transcription. For example, a poly(A) tail and a 5' cap are added to eukaryotic pre-mRNA and introns are removed by the spliceosome. There are also a number of RNA-dependent RNA polymerases that use RNA as their template for synthesis of a new strand of RNA. For instance, a number of RNA viruses (such as poliovirus) use this type of enzyme to replicate their genetic material. Also, RNA-dependent RNA polymerase is part of the RNA interference pathway in many organisms.

Types of RNA

Overview

Messenger RNA (mRNA) is the RNA that carries information from DNA to the ribosome, the sites of protein synthesis (translation) in the cell. The coding sequence of the mRNA determines the amino acid sequence in the protein that is produced. Many RNAs do not code for protein however (about 97% of the transcriptional output is non-protein-coding in eukaryotes).

These so-called non-coding RNAs ("ncRNA") can be encoded by their own genes (RNA genes), but can also derive from mRNA introns. The most prominent examples of non-coding RNAs are transfer RNA (tRNA) and ribosomal RNA (rRNA), both of which are involved in the process of translation. There are also non-coding RNAs involved in gene regulation, RNA processing and other roles. Certain RNAs are able to catalyse chemical reactions such as cutting and ligating other RNA molecules, and the catalysis of peptide bond formation in the ribosome; these are known as ribozymes.

In Translation

Messenger RNA (mRNA) carries information about a protein sequence to the ribosomes, the protein synthesis factories in the cell. It is coded so that every three nucleotides (a codon) correspond to one amino acid. In eukaryotic cells, once precursor mRNA (pre-mRNA) has been transcribed from DNA, it is processed to mature mRNA.

This removes its introns—non-coding sections of the pre-mRNA. The mRNA is then exported from the nucleus to the cytoplasm, where it is bound to ribosomes and translated into its corresponding protein form with the help of tRNA. In prokaryotic cells, which do not have nucleus and cytoplasm compartments, mRNA can bind to ribosomes while it is being transcribed from DNA. After a certain amount of time the message degrades into its component nucleotides with the assistance of ribonucleases.

Transfer RNA (tRNA) is a small RNA chain of about 80 nucleotides that transfers a specific amino acid to a growing polypeptide chain at the ribosomal site of protein synthesis during translation. It has sites for amino acid attachment and an anticodon region for codon recognition that binds to a specific sequence on the messenger RNA chain through hydrogen bonding.

Ribosomal RNA (rRNA) is the catalytic component of the ribosomes. Eukaryotic ribosomes contain four different rRNA molecules: 18S, 5.8S, 28S and 5S rRNA. Three of the rRNA molecules are synthesized in the nucleolus, and one is synthesized elsewhere. In the cytoplasm, ribosomal RNA and protein combine to form a nucleoprotein called a ribosome. The ribosome binds mRNA and carries out protein synthesis. Several ribosomes may be attached to a single mRNA at any time. rRNA is extremely abundant and makes up 80% of the 10 mg/ml RNA found in a typical eukaryotic cytoplasm.

Transfer-messenger RNA (tmRNA) is found in many bacteria and plastids. It tags proteins encoded by mRNAs that lack stop codons for degradation and prevents the ribosome from stalling.

Regulatory RNAs

Several types of RNA can downregulate gene expression by being complementary to a part of an mRNA or a gene's DNA. MicroRNAs (miRNA; 21-22 nt) are found in eukaryotes and act through RNA interference (RNAi), where an effector complex of miRNA and enzymes can break down mRNA to which the miRNA is complementary, block the mRNA from being translated, or accelerate its degradation. While small interfering RNAs (siRNA; 20-25 nt) are often produced by breakdown of viral RNA, there are also endogenous sources of siRNAs.

siRNAs act through RNA interference in a fashion similar to miRNAs. Some miRNAs and siRNAs can cause genes they target to be methylated, thereby decreasing or increasing transcription of those genes. Animals have Piwi-interacting RNAs (piRNA; 29-30 nt) which are active in germline cells and are thought to be a defense against transposons and play a role in gametogenesis.

Many prokaryotes have CRISPR RNAs, a regulatory system similar to RNA interference. Antisense RNAs are widespread; most downregulate a gene, but a few are activators of transcription. One way antisense RNA can act is by binding to an mRNA, forming double-stranded RNA that is enzymatically degraded.

There are many long noncoding RNAs that regulate genes in eukaryotes, one such RNA is Xist, which coats one X chromosome in female mammals and inactivates it. An mRNA may contain regulatory elements itself, such as riboswitches, in the 5' untranslated region or 3' untranslated region; these cis-regulatory elements regulate the activity of that mRNA. The untranslated regions can also contain elements that regulate other genes.

In RNA Processing

Many RNAs are involved in modifying other RNAs. Introns are spliced out of pre-mRNA by spliceosomes, which contain several small nuclear RNAs (snRNA), or the introns can be ribozymes that are spliced by themselves. RNA can also be altered by having its nucleotides modified to other nucleotides than A, C, G and U. In eukaryotes, modifications of RNA nucleotides are generally directed by small nucleolar RNAs (snoRNA; 60-300 nt), found in the nucleolus and cajal bodies. snoRNAs associate with enzymes and guide them to a spot on an RNA by basepairing to that RNA. These enzymes then perform the nucleotide modification. rRNAs and tRNAs are extensively modified, but snRNAs and mRNAs can also be the target of base modification.

RNA Genomes

Like DNA, RNA can carry genetic information. RNA viruses have genomes composed of RNA, and a variety of proteins encoded by that genome. The viral genome is replicated by some of those proteins, while other proteins protect the genome as the virus particle moves to a new host cell. Viroids are another group of pathogens, but they consist only of RNA, do not encode any protein and are replicated by a host plant cell's polymerase.

In Reverse Transcription

Reverse transcribing viruses replicate their genomes by reverse transcribing DNA copies from their RNA; these DNA copies are then transcribed to new RNA. Retrotransposons also spread by copying DNA and RNA from one another, and telomerase contains an RNA that is used as template for building the ends of eukaryotic chromosomes.

Double-stranded RNA

Double-stranded RNA (dsRNA) is RNA with two complementary strands, similar to the DNA found in all cells. dsRNA forms the genetic material of some viruses (double-stranded RNA viruses). Double-stranded RNA such as viral RNA or siRNA can trigger RNA interference in eukaryotes, as well as interferon response in vertebrates.

Key Discoveries in RNA Biology

Research on RNA has led to many important biological discoveries and numerous Nobel Prizes. Nucleic acids were discovered in 1868 by Friedrich Miescher, who called the material 'nuclein' since it was found in the nucleus. It was later discovered that prokaryotic cells, which do not have a nucleus, also contain nucleic acids. The role of RNA in protein synthesis was suspected already in 1939. Severo Ochoa won the 1959 Nobel Prize in Medicine (shared with Arthur Kornberg) after he discovered an enzyme that can synthesize RNA in the laboratory. Ironically, the enzyme discovered by Ochoa (polynucleotide phosphorylase) was later shown to be responsible for RNA degradation, not RNA synthesis.

The sequence of the 77 nucleotides of a yeast tRNA was found by Robert W. Holley in 1965, winning Holley the 1968 Nobel Prize in Medicine (shared with Har Gobind Khorana and Marshall Nirenberg). In 1967, Carl Woese hypothesized that RNA might be catalytic and suggested that the earliest forms of life (self-replicating molecules) could have relied on RNA both to carry genetic information and to catalyze biochemical reactions—an RNA world.

During the early 1970s, retroviruses and reverse transcriptase were discovered, showing for the first time that enzymes could copy RNA into DNA (the opposite of the usual route for transmission of genetic information). For this work, David Baltimore, Renato Dulbecco and Howard Temin were awarded a Nobel Prize in 1975. In 1976, Walter Fiers and his team determined the first complete nucleotide sequence of an RNA virus genome, that of bacteriophage MS2.

In 1977, introns and RNA splicing were discovered in both mammalian viruses and in cellular genes, resulting in a 1993 Nobel to Philip Sharp and Richard Roberts. Catalytic RNA molecules (ribozymes) were discovered in the early 1980s, leading to a 1989 Nobel award to Thomas Cech and Sidney Altman. In 1990 it was found in petunia that introduced genes can silence similar genes of the plant's own, now known to be a result of RNA interference.

At about the same time, 22 nt long RNAs, now called microRNAs, were found to have a role in the development of *C. elegans*. Studies on RNA interference gleaned a Nobel Prize for Andrew Fire and Craig Mello in 2006, and another Nobel was awarded for studies on transcription of RNA to Roger Kornberg in the same year. The discovery of gene regulatory RNAs has led to attempts to develop drugs made of RNA, such as siRNA, to silence genes.

Transcription and Processing of RNA

Gene Expression: Transcription

The majority of genes are expressed as the proteins they encode. The process occurs in two steps:

- Transcription = DNA '! RNA
- Translation = RNA '! protein.

Taken together, they make up the "central dogma" of biology: DNA '! RNA'! protein.

Gene Transcription: DNA '! RNA

DNA serves as the template for the synthesis of RNA much as it does for its own replication.

The Steps

- Some 50 different protein transcription factors bind to promoter sites, usually on the 52 side of the gene to be transcribed.
- An enzyme, an RNA polymerase, binds to the complex of transcription factors.
- Working together, they open the DNA double helix.
- The RNA polymerase proceeds to "read" one strand moving in its 32 '! 52 direction.
- In eukaryotes, this requires — at least for protein-encoding genes — that the nucleosomes in front of the advancing RNA polymerase (RNAP II) be removed. A complex of proteins is responsible for this. The same complex replaces the nucleosomes after the DNA has been transcribed and RNAP II has moved on.
- As the RNA polymerase travels along the DNA strand, it assembles ribonucleotides (supplied as triphosphates, e.g., ATP) into a strand of RNA.

- Each ribonucleotide is inserted into the growing RNA strand following the rules of base pairing. Thus for each C encountered on the DNA strand, a G is inserted in the RNA; for each G, a C; and for each T, an A. However, each A on the DNA guides the insertion of the pyrimidine uracil (U, from uridine triphosphate, UTP). There is no T in RNA.

Quality control: Occasionally RNA polymerase will select and insert an incorrect, mismatched, ribonucleotide. When this occurs in bacteria (and perhaps in all organisms), the enzyme backs up, removes the incorrect nucleotide (and the one preceding it) and tries again. (Described by Zenkin *et al.*, in the 28 July 2006 issue of Science.)

- Synthesis of the RNA proceeds in the 52 '! 32 direction.
- As each nucleoside triphosphate is brought in to add to the 32 end of the growing strand, the two terminal phosphates are removed.
- When transcription is complete, the transcript is released from the polymerase and, shortly thereafter, the polymerase is released from the DNA.

Note that at any place in a DNA molecule, either strand may be serving as the template; that is, some genes "run" one way, some the other (and in a few remarkable cases, the same segment of double helix contains genetic information on both strands!). In all cases, however, RNA polymerase transcribes the DNA strand in its 32 '! 52 direction.

A report in the 4 January 2001 issue of Nature shows that RNA polymerase actually tracks around the double helix of DNA. In vitro, at least, when RNA polymerase is immobilized, it spins the DNA molecule around and around as it moves along the molecule. Whether it is the polymerase or the DNA that does the spinning in vivo remains to be determined.

Types of RNA

Sedimentation pattern produced by high-speed centrifugation of RNA extracted from the precursors of rabbit red blood cells. The discrete bands represent particular classes of RNA.

The transfer RNAs band at about 4S. The ribosomal RNAs of eukaryotes sediment at 5S, 5.8S, 18S, and 28S. (The larger the sedimentation unit, S, the larger the molecule — but not proportionally.) The RNA forming the band at 9S is the messenger RNA for the synthesis of hemoglobin, the major protein synthesized in these cells. In most types of cells, the messenger RNAs are extremely heterogenous, with small amounts distributed from 6S to 25S. Several types of RNA are synthesized in the nucleus of eukaryotic cells. Of particular interest are:

- messenger RNA (mRNA). This will later be translated into a polypeptide.

- ribosomal RNA (rRNA). This will be used in the building of ribosomes: machinery for synthesizing proteins by translating mRNA.
- transfer RNA (tRNA). RNA molecules that carry amino acids to the growing polypeptide.
- small nuclear RNA (snRNA). DNA transcription of the genes for mRNA, rRNA, and tRNA produces large precursor molecules ("primary transcripts") that must be processed within the nucleus to produce the functional molecules for export to the cytosol. Some of these processing steps are mediated by snRNAs.
- small nucleolar RNA (snoRNA). These RNAs within the nucleolus have several functions (described below).
- microRNA (miRNA). These are tiny (~22 nucleotides) RNA molecules that appear to regulate the expression of messenger RNA (mRNA) molecules. [Discussion]
- XIST RNA. This inactivates one of the two X chromosomes in female vertebrates. [Discussion].

Messenger RNA (mRNA)

Messenger RNA comes in a wide range of sizes reflecting the size of the polypeptide it encodes. Most cells produce small amounts of thousands of different mRNA molecules, each to be translated into a peptide needed by the cell.

Many mRNAs are common to most cells, encoding "housekeeping" proteins needed by all cells (e.g. the enzymes of glycolysis). Other mRNAs are specific for only certain types of cells. These encode proteins needed for the function of that particular cell (e.g., the mRNA for hemoglobin in the precursors of red blood cells).

Ribosomal RNA (rRNA)

There are 4 kinds. In eukaryotes, these are ;

- 18S rRNA. One of these molecules, along with some 30 different protein molecules, is used to make the small subunit of the ribosome.
- 28S, 5.8S, and 5S rRNA. One each of these molecules, along with some 45 different proteins, are used to make the large subunit of the ribosome.

The S number given each type of rRNA reflects the rate at which the molecules sediment in the ultracentrifuge. The larger the number, the larger the molecule (but not proportionally).

The 28S, 18S, and 5.8S molecules are produced by the processing of a single primary transcript from a cluster of identical copies of a single gene. The 5S molecules are produced from a different cluster of identical genes.

Transfer RNA (tRNA)

There are some 32 different kinds of tRNA in a typical eukaryotic cell.

- Each is the product of a separate gene.
- They are small (~4S), containing 73-93 nucleotides.
- Many of the bases in the chain pair with each other forming sections of double helix.
- The unpaired regions form 3 loops.
- Each kind of tRNA carries (at its 32 end) one of the 20 amino acids (thus most amino acids have more than one tRNA responsible for them).
- At one loop, 3 unpaired bases form an anticodon.
- Base pairing between the anticodon and the complementary codon on a mRNA molecule brings the correct amino acid into the growing polypeptide chain.

Small Nuclear RNA (snRNA)

Approximately a dozen different genes for snRNAs, each present in multiple copies, have been identified. The snRNAs have various roles in the processing of the other classes of RNA.

For example, several snRNAs are part of the spliceosomes that participate in converting pre-mRNA into mRNA by excising the introns and splicing the exons. [Link down to the discussion of RNA processing.]

Small Nucleolar RNA (snoRNA)

As the name suggests, these small (60–300 nucleotides) RNAs are found in the nucleolus where they are responsible for several functions:

- Some participate in making ribosomes by helping to cut up the large RNA precursor of the 28S, 18S, and 5.8S molecules.
- Others chemically modify many of the nucleotides in rRNA, tRNA, and snRNA molecules, e.g., by adding methyl groups to ribose.
- Some have been implicated in the alternative splicing of pre-mRNA to different forms of mature mRNA.
- One snoRNA serves as the template for the synthesis of telomeres.

In vertebrates, the snoRNAs are made from introns removed during RNA processing.

Noncoding RNA

Only messenger RNA encodes polypeptides. All the other classes of RNA, including types not mentioned here, are thus called noncoding RNA. Much remains to be learned about the function(s) of some of them. But, taken together, noncoding RNAs probably account for two-thirds of the transcription going on in the nucleus.

The RNA Polymerases

The RNA polymerases are huge multi-subunit protein complexes. Three kinds are found in eukaryotes.

- RNA polymerase I (Pol I). It transcribes the rRNA genes for the precursor of the 28S, 18S, and 5.8S molecules (and is the busiest of the RNA polymerases).
- RNA polymerase II (Pol II; also known as RNAP II). It transcribes protein-encoding genes into mRNA (and also the snRNA genes).
- RNA polymerase III (Pol III). It transcribes the 5S rRNA genes and all the tRNA genes.

Two remarkable reports in the 8 June 2001 issue of Science show the structure of Pol II and reveal many details of how it uses the antisense strand of DNA to synthesize a strand of mRNA.

RNA Processing: Pre-mRNA '! mRNA

All the primary transcripts produced in the nucleus must undergo processing steps to produce functional RNA molecules for export to the cytosol. We shall confine ourselves to a view of the steps as they occur in the processing of pre-mRNA to mRNA. Most eukaryotic genes are split into segments. In decoding the open reading frame of a gene for a known protein, one usually encounters periodic stretches of DNA calling for amino acids that do not occur in the actual protein product of that gene. Such stretches of DNA, which get transcribed into RNA but not translated into protein, are called introns. Those stretches of DNA that do code for amino acids in the protein are called exons. Examples:

- The gene for one type of collagen found in chickens is split into 52 separate exons.
- The gene for dystrophin, which is mutated in boys with muscular dystrophy, has 79 exons.
- Even the genes for rRNA and tRNA are split by introns.
- The human genome is estimated to contain some 180,000 exons. With a current estimate of 21,000 genes, the average exon content of our genes is about 9.

In general, introns tend to be much longer than exons. An average eukaryotic exon is only 140 nucleotides long, but one human intron stretches for 480,000 nucleotides! Removal of the introns — and splicing the exons together — are among the essential steps in synthesizing mRNA.

The Steps of RNA Processing

- Synthesis of the cap. This is a modified guanine (G) which is attached to the 52 end of the pre-mRNA as it emerges from RNA polymerase II (RNAP II). The cap
  - o protects the RNA from being degraded by enzymes that degrade RNA from the 52 end;
  - o serves as an assembly point for the proteins needed to recruit the small subunit of the ribosome to begin translation.
- Step-by-step removal of introns present in the pre-mRNA and splicing of the remaining exons. This step takes place as the pre-mRNA continues to emerge from RNAP II.
- Synthesis of the poly(A) tail. This is a stretch of adenine (A) nucleotides. When a special poly(A) attachment site in the pre-mRNA emerges from RNAP II, the transcript is cut there, and the poly(A) tail is attached to the exposed 32 end. This completes the mRNA molecule, which is now ready for export to the cytosol. (The remainder of the transcript is degraded, and the RNA polymerase leaves the DNA).

The cutting and splicing of mRNA must be done with great precision. If even one nucleotide is left over from an intron or one is removed from an exon, the reading frame from that point on will be shifted, producing new codons specifying a totally different sequence of amino acids from that point to the end of the molecule (which often ends prematurely anyway when the shifted reading frame generates a STOP codon).

The removal of introns and splicing of exons is done by spliceosomes. These are a complexes of 5 snRNA molecules and some 145 different proteins.

The introns in most pre-mRNAs begin with a GU and end with an AG. Presumably these short sequences assist in guiding the spliceosome.

Visual Evidence

The upper image is an electron micrograph of a mRNA-DNA hybrid molecule formed by mixing the messenger RNA (mRNA) from a clone of antibody-secreting cells with single-stranded DNA from the same kind of cells. The bar represents the length of 1000 bases.

The lower diagram is an interpretation of the micrograph. The solid line represents the DNA; the dotted line the mRNA. The loops (I_A, I_B, etc.) represent the introns that separate the exons encoding the domains of an antibody heavy chain:

- V = variable region
- E_1 = first constant region (C_H1) domain

- E_H = hinge region
- E_2 and E_3 = the nucleotides encoding the two C-terminal domains (C_H2 and C_H3).

The unhybridized portion of the mRNA is its poly(A) tail.

Alternative Splicing

The processing of pre-mRNA for many proteins proceeds along various paths in different cells or under different conditions. For example, early in the differentiation of a B cell (a lymphocyte that synthesizes an antibody) the cell first uses an exon that encodes a transmembrane domain that causes the molecule to be retained at the cell surface. Later, the B cell switches to using a different exon whose domain enables the protein to be secreted from the cell as a circulating antibody molecule.

Alternative splicing provides a mechanism for producing a wide variety of proteins from a small number of genes. While we humans may turn out to have only some 20 thousand genes, we probably make at least 10 times that number of different proteins. It is now estimated that 92–94% of our genes produce pre-mRNAs that are alternatively-spliced. There is evidence that the pattern of alternative splicing differs consistently in different tissues and so must be regulated. But whether all the products are functional or that many are simply the outcome of an error-prone process remains to be seen.

Alternative splicing not only provides different proteins from a single gene but also different 3' UTRs and 5' UTRs. Although not translated into protein, these untranslated regions contain signals that, for example, dictate where in the cell that protein will accumulate. Two examples:

- The 3' UTR of the *bicoid* gene in Drosophila directs the mRNA to the anterior of the embryo;
- the same region in the *VegT* gene of Xenopus directs its mRNA to the vegetal pole of the embryo.

One of the most dramatic examples of alternative splicing is the *Dscam* gene in Drosophila. This single gene contains some 116 exons of which 17 are retained in the final mRNA. Some exons are always included; others are selected from an array. Theoretically this system is able to produce 38,016 different proteins. And, in fact, over 18,000 different ones have been found in Drosophila hemolymph.

These Dscam proteins are used to establish a unique identity for each neuron. It works like this. Each developing neuron draws upon the pool of thousands of possible different mRNAs to synthesize a dozen or so of them. Which ones are selected appears to be simply a matter of chance, but because of the size of the pool, each neuron will most likely end up with a unique set of a dozen or so Dscam proteins. As each developing neuron in the central nervous system sprouts dendrites and an axon, these express its unique collection of Dscam proteins. If the extensions of a single neuron should meet each other in the tangled web that is the hallmark of nervous tissue, they are repelled. In this way, thousands of different neurons can coexist in intimate contact without the danger of nonfunctional contacts between the various extensions of the same neuron.

We also have a *DSCAM* gene which is located on chromosome 21. When present in 3 copies, it is probably responsible for some of the features of Down syndrome (and accounts for its full name: Down Syndrome Cell Adhesion Molecule). Perhaps the incredible diversity of synaptic junctions in the mammalian c.n.s. ($\sim 10^{14}$) is mediated by alternative splicing of a limited number of gene transcripts. So, whether a particular segment of RNA will be retained as an exon or excised as an intron can vary under different circumstances, such as ;

- what type of cell the gene is in;
- what stage of differentiation that cell is passing through;
- what extracellular signals that cell is receiving.

Clearly the switching to an alternate splicing pathway must be closely regulated.

Trans-splicing

Most genes are transcribed and their transcripts processed as described above. RNA polymerase travels down a single strand of a singe gene locus to form pre-mRNA that is processed (including removal of introns) to form the mature mRNA. But there are exceptions. A number of cases have been found where two different precursor transcripts have been spliced together to form the final RNA molecule. The phenomenon is called *trans*-splicing.

Examples: synthesis of a single RNA molecule by splicing together transcripts from ;

- loci located far apart on the same chromosome;
- opposite strands of the same gene locus;
- the two alleles of the gene on their separate (homologous) chromosomes.

The biological importance of these *trans*-spliced transcripts is still unknown for most of them.

Why Split Genes?

Perhaps during evolution, eukaryotic genes have been assembled from smaller, primitive genes - today's exons. Some proteins, like the antibodies mentioned in the previous section, are organized in a set of separate sections or domains each with a special function to perform in the complete molecule. Each domain is encoded by a separate exon. Having the different functional parts of the antibody molecule encoded by separate exons makes it possible to use these units in different combinations. Thus a set of exons in the genome may be the genetic equivalent of the various modular pieces in a box of "Lego" for children to assemble in whatever forms they wish. But the boundaries of other exons do not seem to correspond to domain boundaries of the protein. So perhaps the major benefit of split genes is simply the opportunity they provide for making many different proteins from a single gene through alternative splicing.

Summary

Gene expression occurs in two steps:

- transcription of the information encoded in DNA into a molecule of RNA (described here) and
- translation of the information encoded in the nucleotides of mRNA into a defined sequence of amino acids in a protein (discussed in Gene Translation: RNA → Protein).

Messenger RNA

Messenger RNA (mRNA) is a molecule of RNA encoding a chemical "blueprint" for a protein product. mRNA is transcribed from a DNA template, and carries coding information to the sites of protein synthesis: the ribosomes.

Here, the nucleic acid polymer is translated into a polymer of amino acids: a protein. In mRNA as in DNA, genetic information is encoded in the sequence of nucleotides arranged into codons consisting of three bases each. Each codon encodes for a specific amino acid, except the stop codons that terminate protein synthesis. This process requires two other types of RNA: transfer RNA (tRNA) mediates recognition of the codon and provides the corresponding amino acid, while ribosomal RNA (rRNA) is the central component of the ribosome's protein manufacturing machinery.

Synthesis, Processing, and Function

The brief existence of an mRNA molecule begins with transcription and ultimately ends in degradation. During its life, an mRNA molecule may also be processed, edited, and transported prior to translation. Eukaryotic mRNA molecules often require extensive processing and transport, while prokaryotic molecules do not.

Transcription

During transcription, RNA polymerase makes a copy of a gene from the DNA to mRNA as needed. This process is similar in eukaryotes and prokaryotes.

One notable difference, however, is that prokaryotic RNA polymerase associates with mRNA processing enzymes during transcription so that processing can proceed quickly after the start of transcription. The short-lived, unprocessed or partially processed, product is termed *pre-mRNA*; once completely processed, it is termed *mature mRNA*.

Eukaryotic Pre-mRNA Processing

Processing of mRNA differs greatly among eukaryotes, bacteria and archea. Non-eukaryotic mRNA is essentially mature upon transcription and requires no processing, except in rare cases. Eukaryotic pre-mRNA, however, requires extensive processing.

5' Cap Addition

A *5' cap* (also termed an RNA cap, an RNA 7-methylguanosine cap or an RNA m^7G cap) is a modified guanine nucleotide that has been added to the "front" or 5' end of a eukaryotic messenger RNA shortly after the start of transcription. The 5' cap consists of a terminal 7-methylguanosine residue which is linked through a 5'-5'-triphosphate bond to the first transcribed nucleotide.

Its presence is critical for recognition by the ribosome and protection from RNases. Cap addition is coupled to transcription, and occurs co-transcriptionally, such that each influences the other. Shortly after the start of transcription, the 5' end of the mRNA being synthesized is bound by a cap-synthesizing complex associated with RNA polymerase. This enzymatic complex catalyzes the chemical reactions that are required for mRNA capping. Synthesis proceeds as a multi-step biochemical reaction.

Splicing

Splicing is the process by which pre-mRNA is modified to remove certain stretches of non-coding sequences called introns; the stretches that remain include protein-coding sequences and are called exons. Sometimes pre-mRNA messages may be spliced in several different ways, allowing a single gene to encode multiple proteins. This process is called alternative splicing. Splicing is usually performed by an RNA-protein complex called the spliceosome, but some RNA molecules are also capable of catalyzing their own splicing.

Ribozyme

A ribozyme (from ribonucleic acid enzyme, also called RNA enzyme or catalytic RNA) is an RNA molecule possessing a well defined tertiary structure that enables it to catalyze a chemical reaction. Many natural ribozymes catalyze either the hydrolysis of one of their own phosphodiester bonds (self-cleaving ribozymes), or the hydrolysis of bonds in other RNAs, but they have also been found to catalyze the aminotransferase activity of the ribosome. Investigators studying the origin of life have produced ribozymes in the laboratory that are capable of catalyzing their own synthesis under very specific conditions, such as an RNA polymerase ribozyme. Mutagenesis and selection has been performed resulting in isolation of improved variants of the "Round-18" polymerase ribozyme from 2001. "B6.61" is able to add up to 20 nucleotides to a primer template in 24 hours, until it decomposes by hydrolysis of its phosphodiester bonds. Some ribozymes may play an important role as therapeutic agents, as enzymes which tailor defined RNA sequences, as biosensors, and for applications in functional genomics and gene discovery.

Discovery

Before the discovery of ribozymes, enzymes, which are defined as catalytic proteins, were the only known biological catalysts. In 1967, Carl Woese, Francis Crick, and Leslie Orgel were the first to suggest that RNA could act as a catalyst. This idea was based upon the discovery that RNA can form complex secondary structures. The first ribozymes were discovered in the 1980s by Thomas R. Cech, who was studying RNA splicing in the ciliated protozoan *Tetrahymena thermophila* and Sidney Altman, who was working on the bacterial RNase P complex.

These ribozymes were found in the intron of an RNA transcript, which removed itself from the transcript, as well as in the RNA component of the RNase P complex, which is involved in the maturation of pre-tRNAs. In 1989, Thomas R. Cech and Sidney Altman won the Nobel Prize in chemistry for their "discovery of catalytic properties of RNA." The term *ribozyme* was first introduced by Kelly Kruger *et al.* in 1982 in a paper published in *Cell*.

It had been a firmly established belief in biology that catalysis was reserved for proteins. In retrospect, catalytic RNA makes a lot of sense. This is based on the old question regarding the origin of life: Which comes first, enzymes that do the work of the cell or nucleic acids that carry the information required to produce the enzymes? Nucleic acids as catalysts circumvents this problem.

In the 1970s Thomas Cech, at the University of Colorado at Boulder, was studying the excision of introns in a ribosomal RNA gene in *Tetrahymena thermophila*. While trying to purify the enzyme responsible for splicing reaction, he found that intron could be spliced out in the absence of any added cell extract. As much as they tried, Cech and his colleagues could not identify any protein associated with the splicing reaction.

After much work, Cech proposed that the intron sequence portion of the RNA could break and reform phosphodiester bonds. At about the same time, Sidney Altman, who is a Professor at Yale University, was studying the way tRNA molecules are processed in the cell when he and his colleagues isolated an enzyme called RNase-P, which is responsible for conversion of a precursor tRNA into the active tRNA. Much to their surprise, they found that RNase-P contained RNA in addition to protein and that RNA was an essential component of the active enzyme.

This was such a foreign idea that they had difficulty publishing their findings. The following year, Altman demonstrated that RNA can act as a catalyst by showing that the RNase-P RNA subunit could catalyze the cleavage of precursor tRNA into active tRNA in the absence of any protein component.

Since Cech's and Altman's discovery, other investigators have discovered other examples of self-cleaving RNA or catalytic RNA molecules. Many ribozymes have either a hairpin – or hammerhead – shaped active center and a unique secondary structure that allows them to cleave other RNA molecules at specific sequences.

It is now possible to make ribozymes that will specifically cleave any RNA molecule. These RNA catalysts may have pharmaceutical applications. For example, a ribozyme has been designed to cleave the RNA of HIV. If such a ribozyme was made by a cell, all incoming virus particles would have their RNA genome cleaved by the ribozyme, which would prevent infection.

Activity

Although most ribozymes are quite rare in the cell, their roles are sometimes essential to life. For example, the functional part of the ribosome, the molecular machine that translates RNA into proteins, is fundamentally a ribozyme, composed of RNA tertiary structural motifs that are often coordinated to metal ions such as Mg^{2+} as cofactors.

There is no requirement for divalent cations in a five-nucleotide RNA that can catalyze *trans*-phenylalanation of a four-nucleotide substrate which has three base complementary sequence with the catalyst. The catalyst and substrate were devised by truncation of the C3 ribozyme.

RNA can also act as a hereditary molecule, which encouraged Walter Gilbert to propose that in the distant past, the cell used RNA as both the genetic material and the structural and catalytic molecule, rather than dividing these functions between DNA and protein as they are today. This hypothesis became known as the "RNA world hypothesis" of the origin of life.

If ribozymes were the first molecular machines used by early life, then today's remaining ribozymes — such as the ribosome machinery — could be considered living fossils of a life based primarily on nucleic acids. A recent test-tube study of prion folding suggests that an RNA may catalyze the pathological protein conformation in the manner of a chaperone enzyme.

Known Ribozymes

Naturally occurring ribozymes include:

- Peptidyl transferase 23S rRNA
- RNase P
- Group I and Group II introns
- GIR1 branching ribozyme
- Leadzyme - Although initially created *in vitro*, natural examples have been found
- Hairpin ribozyme
- Hammerhead ribozyme
- HDV ribozyme
- Mammalian CPEB3 ribozyme
- VS ribozyme
- *glmS* ribozyme
- CoTC ribozyme.

Artificial Ribozymes

Since the discovery of ribozymes that exist in living organisms, there has been interest in the study of new synthetic ribozymes made in the laboratory. For example, artificially-produced self-cleaving RNAs that have good enzymatic activity have been produced. Tang and Breaker isolated self-cleaving RNAs by in vitro selection of RNAs originating from random-sequence RNAs. Some of the synthetic ribozymes that were produced had novel structures, while some were similar to the naturally occurring hammerhead ribozyme.

The techniques used to discover artificial ribozymes involve Darwinian evolution. This approach takes advantage of RNA's dual nature as both a catalyst and an informational polymer, making it easy for an investigator to produce vast populations of RNA catalysts using polymerase enzymes. The ribozymes are mutated by reverse transcribing them with reverse transcriptase into various cDNA and amplified with mutagenic PCR. The selection parameters in these experiments often differ. One approach for selecting a ligase ribozyme involves using biotin tags, which are covalently linked to the substrate. If a molecule possesses the desired ligase activity, a streptavidin matrix can be used to recover the active molecules.

Lincoln and Joyce developed an RNA enzyme system capable of self replication in about an hour. By utilizing molecular competition (*in vitro* evolution) of a candidate enzyme mixture, a pair of RNA enzymes emerged, in which each synthesizes the other from synthetic oligonucleotides, with no protein present.

Applications

A type of synthetic ribozyme directed against HIV RNA called gene shears has been developed and has entered clinical testing for HIV infection.

Editing

In some instances, an mRNA will be edited, changing the nucleotide composition of that mRNA. An example in humans is the apolipoprotein B mRNA, which is edited in some tissues, but not others. The editing creates an early stop codon, which upon translation, produces a shorter protein.

Polyadenylation

Polyadenylation is the addition of a poly(A) tail to an RNA molecule. The poly(A) tail consists of multiple adenosine monophosphates; in other words, it is a stretch of RNA which only has adenine bases. In eukaryotes, polyadenylation is part of the process that produces mature messenger RNA (mRNA) for translation. It therefore forms part of the larger process of gene expression.

The process of polyadenylation begins as the transcription of a gene finishes. The 3'-most segment of the newly-made RNA is first cleaved off by a set of proteins; these proteins then synthesize the poly(A) tail at the RNA's 3' end. In some genes these proteins may add a poly(A) tail at any one of several possible sites, polyadenylation can therefore produce more than one transcript from a single gene, similar to alternative splicing.

The poly(A) tail is important for the nuclear export, translation and stability of mRNA. The tail is shortened over time and when it is short enough, the mRNA is enzymatically degraded. However, in a few cell types, mRNAs with short poly(A) tails are stored for later activation by re-polyadenylation in the cytosol. In contrast, when polyadenylation occurs in bacteria, it promotes RNA degradation. This is also sometimes the case for eukaryotic non-coding RNAs. The wide distribution of polyadenylation among living organisms indicates that this process evolved early in the history of life on Earth.

Primer on RNA

RNAs are a type of large biological molecules, whose individual building blocks are called nucleotides. The name *poly(A) tail* (for polyadenylic acid tail) reflects the way RNA nucleotides are abbreviated, with a letter for the base the nucleotide contains (A for adenine, C for cytosine, G for guanine and U for uracil). RNAs are produced (*transcribed*) from a DNA template. By convention, RNA sequences are written in a 5' to 3' direction. The 5' end is the part of the RNA molecule that is transcribed first, and the 3' end is transcribed last. The 3' end is also where the poly(A) tail is found on polyadenylated RNAs.

Messenger RNA (mRNA) is RNA that has a coding region that acts as a template for protein synthesis (*translation*). The rest of the mRNA, the *untranslated regions*, tune how active the mRNA is. There are also many RNAs that are not translated, called non-coding RNAs. Like the untranslated regions, many of these non-coding RNAs have regulatory roles.

Nuclear Polyadenylation

Function

In nuclear polyadenylation, a poly(A) tail is added to an RNA at the end of transcription. On mRNAs, the poly(A) tail protects the mRNA molecule from enzymatic degradation in the cytoplasm and aids in transcription termination, export of the mRNA from the nucleus, and translation. Almost all eukaryotic mRNAs are polyadenylated, with the exception of animal replication-dependent histone mRNAs. These are the only mRNAs in eukaryotes that lack a poly(A) tail, ending instead in a stem-loop structure followed by a purine-rich sequence, termed histone downstream element, that directs where the RNA is cut so that the 3' end of the histone mRNA is formed.

Many eukaryotic non-coding RNAs are always polyadenylated at the end of transcription. There are small RNAs where the poly(A) tail is only seen in intermediary forms and not in the mature RNA as the ends are removed during processing, notably microRNAs. But for many long noncoding RNAs – a seemingly large group of regulatory RNAs that for example includes the RNA Xist which mediates X chromosome inactivation – a poly(A) tail is part of the mature RNA.

Mechanism

The polyadenylation machinery in the nucleus of eukaryotes works on products of RNA polymerase II, such as precursor mRNA. Here, a multi-protein complex cleaves the 3'-most part of a newly produced RNA and polyadenylates the end produced by this cleavage. The cleavage is catalysed by the enzyme CPSF and occurs 10–30 nucleotides downstream of its binding site. This site is often the sequence AAUAAA on the RNA, but variants of it exist that bind more weakly to CPSF.

Two other proteins add specificity to the binding to an RNA: CstF and CFI. CstF binds to a GU-rich region further downstream of CPSF's site. CFI recognises a third site on the RNA (a set of UGUAA sequences in mammals) and can recruit CPSF even if the AAUAAA sequence is missing. The polyadenylation signal – the sequence motif recognised by the RNA cleavage complex – varies between groups of eukaryotes. Most human polyadenylation sites contain the AAUAAA sequence, but this sequence is less common in plants and fungi.

The RNA is cleaved right after transcription, as CstF also binds to RNA polymerase II. Cleavage also involves the protein CFII, though it is unknown how. The cleavage site associated with a polyadenylation signal can vary up to some 50 nucleotides. When the RNA is cleaved, polyadenylation starts, catalysed by polyadenylate polymerase. Polyadenylate polymerase builds the poly(A) tail by adding adenosine monophosphate units from adenosine triphosphate to the RNA, cleaving off pyrophosphate.

Another protein, PAB2, binds to the new, short poly(A) tail and increases the affinity of polyadenylate polymerase for the RNA. When the poly(A) tail is approximately 250 nucleotides long the enzyme can no longer bind to CPSF and polyadenylation stops, thus determining the length of the poly(A) tail. CPSF is in contact with RNA polymerase II, allowing it to tell the polymerase to terminate transcription. The polyadenylation machinery is also physically linked to the spliceosome, a complex that removes introns from RNAs.

Downstream Effects

The poly(A) tail acts as the binding site for poly(A)-binding protein. Poly(A)-binding protein promotes export from the nucleus and translation, and inhibits degradation. This protein binds to the poly(A) tail prior to mRNA export from the nucleus and in yeast also recruits poly(A) nuclease, an enzyme that shortens the poly(A) tail and allows the export of the mRNA. Poly(A)-binding protein is exported to the cytoplasm with the RNA. mRNAs that are not exported are degraded by the exosome. Poly(A)-binding protein also can bind to, and thus recruit, several proteins that affect translation, one of these is initiation factor-4G which in turn recruits the 40S ribosomal subunit. However, a poly(A) tail is not required for the translation of all mRNAs.

Deadenylation

In eukaryotic somatic cells, the poly(A) tail of most mRNAs in the cytoplasm gradually get shorter, and mRNAs with shorter poly(A) tail are translated less and degraded sooner. However, it can take many hours before an mRNA is degraded. This deadenylation and degradation process can be accelerated by microRNAs complementary to the 3' untranslated region of an mRNA. In immature egg cells, mRNAs with shortened poly(A) tails are not degraded, but are instead stored without being translated. They are then activated by cytoplasmic polyadenylation after fertilisation, during egg activation.

In animals, poly(A) ribonuclease (PARN) can bind to the 5' cap and remove nucleotides from the poly(A) tail. The level of access to the 5' cap and poly(A) tail is important in controlling how soon the mRNA is degraded. PARN deadenylates less if the RNA is bound by the initiation factors 4E (at the 5' cap) and 4G (at the poly(A) tail), so this is why translation reduces deadenylation. The rate of deadenylation may also be regulated by RNA-binding proteins.

Once the poly(A) tail is removed, the decapping complex removes the 5' cap, leading to a degradation of the RNA. Several other enzymes that seem to be involved in deadenylation have been identified in yeast.

Alternative Polyadenylation

Many protein-coding genes have more than one polyadenylation site, so a gene can code for several mRNAs that differ in their 3' end. Since alternative polyadenylation changes the length of the 3' untranslated region, it can change which binding sites for microRNAs the 3' untranslated region contains. MicroRNAs tend to repress translation and promote degradation of the mRNAs they bind to, although there are examples of microRNAs that stabilise transcripts. Alternative polyadenylation can also shorten the coding region, thus making the mRNA code for a different protein, but this is much less common than just shortening the 3' untranslated region.

The choice of poly(A) site can be influenced by extracellular stimuli and depends on the expression of the proteins that take part in polyadenylation. For example, the expression of CstF-64, a subunit of cleavage stimulatory factor (CstF), increases in macrophages in response to lipopolysaccharides (a group of bacterial compounds that trigger an immune response). This results in the selection of weak poly(A) sites and thus shorter transcripts.

This removes regulatory elements in the 3' untranslated regions of mRNAs for defense-related products like lysozyme and TNF-α. These mRNAs then have longer half-lives and produce more of these proteins. RNA-binding proteins other than those in the polyadenylation machinery can also affect whether a polyadenyation site is used,as can DNA methylation near the polyadenylation signal.

Cytoplasmic Polyadenylation

There is polyadenylation in the cytosol of some animal cell types, namely in the germ line, during early embryogenesis and in post-synaptic sites of nerve cells. This lengthens the poly(A) tail of an mRNA with a shortened poly(A) tail, so that the mRNA will be translated. These shortened poly(A) tails are often less than 20 nucleotides, and are lengthened to around 80–150 nucleotides.

In the early mouse embryo, cytoplasmic polyadenylation of maternal RNAs from the egg cell allows the cell to survive and grow even though transcription does not start until the middle of the 2-cell stage (4-cell stage in human). In the brain, cytoplasmic polyadenylation is active during learning and could play a role in long-term potentiation, which is the strengthening of the signal transmission from a nerve cell to another in response to nerve impulses and is important for learning and memory formation.

Cytoplasmic polyadenylation requires the RNA-binding proteins CPSF and CPEB, and can involve other RNA-binding proteins like Pumilio. Depending on the cell type, the polymerase can be the same type of polyadenylate polymerase (PAP) that is used in the nuclear process, or the cytoplasmic polymerase GLD-2.

Tagging for Degradation in Eukaryotes

For many non-coding RNAs, including tRNA, rRNA, snRNA and snoRNA, polyadenylation is a way of marking the RNA for degradation, in at least yeast. This polyadenylation is done in the nucleus by the TRAMP complex, which adds a tail that is around 40 nucleotides long to the 3' end. The RNA is then degraded by the exosome. Poly(A) tails have also been found on human rRNA fragments, both the form of homopolymeric (A only) and heterpolymeric (mostly A) tails.

In Prokaryotes and Organelles

In many bacteria, both mRNAs and non-coding RNAs can be polyadenylated. This poly(A) tail promotes degradation by the degradosome, which contains two RNA-degrading enzymes: polynucleotide phosphorylase and RNase E. Polynucleotide phosphorylase binds to the 3' end of RNAs and the 3' extension provided by the poly(A) tail allows it to bind to the RNAs whose secondary structure would otherwise block the 3' end. Successive rounds of polyadenylation and degradation of the 3' end by polynucleotide phosphorylase allows the degradosome to overcome these secondary structures.

The poly(A) tail can also recruit RNases that cut the RNA in two. These bacterial poly(A) tails are about 30 nucleotides long. In as different groups as animals and trypanosomes, the mitochondria contain both stabilising and destabilising poly(A) tails. Destabilising polyadenylation targets both mRNA and noncoding RNAs.

The poly(A) tails are 43 nucleotides long on average. The stabilising ones start at the stop codon, and without them the stop codon (UAA) is not complete as the genome only encodes the U or UA part. Plant mitochondria only have destabilising polyadenylation, and yeast mitochondria have no polyadenylation at all.

While many bacteria and mitochondria have polyadenylate polymerases, they also have another type of polyadenylation, performed by polynucleotide phosphorylase itself. This enzyme is found in bacteria, mitochondria, plastids and as a constituent of the archeal exosome (in those archaea that have an exosome). It can synthesise a 3' extension where the vast majority of the bases are adenines. Like in bacteria, polyadenylation by polynucleotide phosphorylase promotes degradation of the RNA in plastids and likely also archaea.

Evolution

Although polyadenylation is seen in almost all organisms, it is not universal. However, the wide distribution of this modification and the fact that it is present in organisms from all three domains of life implies that the last universal common ancestor of all living organisms probably had some form of polyadenylation system. A few organisms do not polyadenylate mRNA, which implies that they have lost their polyadenylation machineries during evolution. Although no examples of eukaryotes that lack polyadenylation are known, mRNAs from the bacterium *Mycoplasma gallisepticum* and the salt-tolerant archaean *Haloferax volcanii* lack this modification.

The most ancient polyadenylating enzyme is polynucleotide phosphorylase. This enzyme is part of both the bacterial degradosome and archaeal exosome, two closely related complexes that recycle RNA into nucleotides. This enzyme degrades RNA by attacking the bond between the 3'-most nucleotides with a phosphate, breaking off a diphosphate nucleotide. This reaction is reversible, and so the enzyme can also extend RNA with more nucleotides.

The heteropolymeric tail added by polynucleotide phosphorylase is very rich in adenine. The choice of adenine is most likely the result of higher ADP concentrations than other nucleotides as a result of using ATP as an energy currency, making it more likely to be incorporated in this tail in early lifeforms. It has been suggested that the involvement of adenine-rich tails in RNA degradation prompted the later evolution of polyadenylate polymerases (the enzymes that produce poly(A) tails with no other nucleotides in them).

Polyadenylate polymerases are not as ancient. They have separately evolved in both bacteria and eukaryotes from CCA-adding enzyme, which is the enzyme that completes the 3' ends of tRNAs. Its catalytic domain is homologous to that of other polymerases. Likely, the horizontal transfer of bacterial CCA-adding enzyme to eukaryotes allowed the archaeal-like CCA-adding enzyme to switch function to a poly(A) polymerase. Some lineages, like archaea and cyanobacteria, never evolved a polyadenylate polymerase.

History

Polyadenylation was first identified in 1960 as an enzymatic activity in extracts made from cell nuclei that could polymerise ATP, but not ADP, into polyadenine. Although identified in many types of cells, this activity had no known function until 1971, when poly(A) sequences were found in mRNAs. Initially, the only function of these sequences was thought to be protection of the 3' end of the RNA from nucleases, but later the specific roles of polyadenylation in nuclear export and translation were identified. The polymerases responsible for polyadenylation were first purified and characterized in the 1960s and 1970s, but the large number of accessory proteins that control this process were only discovered in the early 1990s. Polyadenylation is the covalent linkage of a polyadenylyl moiety to a messenger RNA molecule.

In eukaryotic organisms, most messenger RNA (mRNA) molecules are polyadenylated at the 3' end. The poly(A) tail and the protein bound to it aid in protecting mRNA from degradation by exonucleases. Polyadenylation is also important for transcription termination, export of the mRNA from the nucleus, and translation. mRNA can also be polyadenylated in prokaryotic organisms, where poly(A) tails act to facilitate, rather than impede, exonucleolytic degradation. Polyadenylation occurs during and immediately after transcription of DNA into RNA.

After transcription has been terminated, the mRNA chain is cleaved through the action of an endonuclease complex associated with RNA polymerase. After the mRNA has been cleaved, around 250 adenosine residues are added to the free 3' end at the cleavage site. This reaction is catalyzed by polyadenylate polymerase. Just as in alternative splicing, there can be more than one polyadenylation variant of a mRNA.

Transport

Another difference between eukaryotes and prokaryotes is mRNA transport. Because eukaryotic transcription and translation is compartmentally separated, eukaryotic mRNAs must be exported from the nucleus to the cytoplasm. Mature mRNAs are recognized by their processed modifications and then exported through the nuclear pore. In neurons mRNA must be transported from the soma to the dendrites where local translation occurs in response to external stimuli. Many messages are marked with so-called "zip codes" which targets their transport to a specific location.

Translation

Because prokaryotic mRNA does not need to be processed or transported, translation by the ribosome can begin immediately after the end of transcription. Therefore, it can be said that prokaryotic translation is *coupled* to transcription and occurs *co-transcriptionally*.

Eukaryotic mRNA that has been processed and transported to the cytoplasm (i.e. mature mRNA) can then be translated by the ribosome. Translation may occur at ribosomes free-floating in the cytoplasm, or directed to the endoplasmic reticulum by the signal recognition particle. Therefore, unlike in prokaryotes, eukaryotic translation *is not* directly coupled to transcription.

5' Cap Structure

The 5' cap is found on the 5' end of an mRNA molecule and consists of a guanine nucleotide connected to the mRNA via an unusual 5' to 5' triphosphate linkage. This guanosine is methylated on the 7 position directly after capping *in vivo* by a methyl transferase. It is referred to as a 7-methylguanosine cap, abbreviated m^7G.

Further modifications include the possible methylation of the 2' hydroxy-groups of the first 2 ribose sugars of the 5' end of the mRNA. The methylation of both 2' hydroxy-groups is shown on the diagram. The 5' cap looks like the 3' end of an RNA molecule (the 5' carbon of the cap ribose is bonded, and the 3' unbonded). This provides significant resistance to 5' exonucleases.

Capping Process

The starting point is the unaltered 5' end of an RNA molecule. This features a final nucleotide followed by three phosphate groups attached to the 5' carbon.

1. One of the terminal phosphate groups is removed (by RNA terminal phosphatase), leaving two terminal phosphates.
2. GTP is added to the terminal phosphates (by a guanylyl transferase), losing two phosphate groups (from the GTP) in the process. This results in the 5' to 5' triphosphate linkage.
3. The 7-nitrogen of guanine is methylated (by a methyl transferase).
4. Other methyltransferases are optionally used to carry out methylation of 5' proximal nucleotides.

5' Capping Targeting

The Capping Enzyme Complex (CEC) required for capping is found bound to the RNA polymerase II before transcription starts. As soon as the 5' end of the new transcript emerges the enzymes transfer to it and begin the capping process (this is a similar kind of mechanism to ensure capping as for polyadenylation). The enzymes for capping can only bind to RNA polymerase II ensuring specificity to only these transcripts, which are almost entirely mRNA.

5' Cap Function

The 5' cap has 4 main functions:

1. Regulation of nuclear export.
2. Prevention of degradation by exonucleases.
3. Promotion of translation.
4. Promotion of 5' proximal intron excision.

Nuclear export of RNA is regulated by the Cap binding complex (CBC), which binds exclusively to capped RNA. The CBC is then recognized by the nuclear pore complex and exported. Once in the cytoplasm after the pioneer round of translation, the CBC is replaced by the translation factors eIF-4E and eIF-4G. This complex is then recognized by other translation initiation machinery including the ribosome. Cap prevents 5' degradation in two ways. First, degradation of the mRNA by 5' exonucleases is prevented (as mentioned above) by functionally looking like a 3' end. Second, the CBC complex and the eIF-4E/eIF-4G block the access of decapping enzymes to the cap. This increases the half-life of the mRNA, essential in eukaryotes as the export process takes significant time.

Decapping of an mRNA is catalyzed by the decapping complex made up of at least Dcp1 and Dcp2, which must compete with eIF-4E to bind the cap. Thus the 5' cap is a marker of an actively translating mRNA and is used by cells to regulate mRNA half-lives in response to new stimuli. Undesirable mRNAs are sent to P-bodies for temporary storage or decapping, the details of which are still being resolved. The mechanism of 5' proximal intron excision promotion is not well understood, but the 5' cap appears to loop around and interact with the spliceosome in the splicing process, promoting intron excision.

Messenger RNA Decapping

The process of messenger RNA decapping consists of hydrolysis of the 5' cap structure on the RNA exposing a 5' monophosphate. This 5' monophosphate is a substrate for the exonuclease Xrn1 and the message is quickly destroyed. There are many situations which may lead to the removal of the cap, some of which are discussed below.

Translation and Decay

Inside cells there is a balance between the processes of translation and mRNA decay. Messages which are being actively translated are bound by polysomes and the initiation factors eIF-4E and eIF-4G. This blocks access to the cap by the decapping enzyme DCP2 and protects the message. In nutrient-starvation conditions or viral infection, translation may be compromised and decapping is stimulated. This balance is reflected in the size and abundance of the cytoplasmic structures known as P-bodies

Specific Decay Pathways

A number of specific decay pathways exist that recognize aberrant messages and promote their decapping. Nonsense mediated decay recognizes premature stop codons and promotes decapping as well as decay by the exosome. Certain classes of miRNA have also been shown to stimulate decapping.

Coding Regions

The coding region of a gene is that portion of a gene's DNA or RNA, composed of exons, that codes for protein. The region is bounded nearer the

5' end by a start codon and nearer the 3' end with a stop codon. The coding region in mRNA is bounded by the five prime untranslated region and the three prime untranslated region, which are also parts of the exons.

The coding region of an organism is sum total of the organism's genome that is composed of gene coding regions.

Coding regions are composed of codons, which are decoded and translated (in eukaryotes usually into one and in prokaryotes usually into several) proteins by the ribosome.

Coding regions begin with the start codon and end with a stop codon. Generally, the start codon is an AUG triplet and the stop codon is UAA, UAG, or UGA. The coding regions tend to be stabilised by internal base pairs, this impedes degradation. In addition to being protein-coding, portions of coding regions may serve as regulatory sequences in the pre-mRNA as exonic splicing enhancers or exonic splicing silencers.

Open Reading Frame

In molecular genetics, an open reading frame (ORF) is a DNA sequence that does not contain a stop codon in a given reading frame.

Significance

One common use of open reading frames are as one piece of evidence to assist in gene prediction. Long ORFs are often used, along with other evidence, to initially identify candidate protein coding regions in a DNA sequence.

The presence of an ORF does not necessarily mean that the region is ever translated. For example in a randomly generated DNA sequence with an equal percentage of each nucleotide, a stop-codon would be expected once every 21 codons. A simple gene prediction algorithm for prokaryotes might look for a start codon followed by an open reading frame that is long enough to encode a typical protein, where the codon usage of that region matches the frequency characteristic for the given organism's coding regions. By itself even a long open reading frame is not conclusive evidence for the presence of a gene.

Example

If a portion of a genome has been sequenced (e.g. 5'-ATCTAAAATGGGTGCC-3'), ORFs can be located by examining each of the three possible reading frames on each strand. In this sequence two out of three possible reading frames are entirely open, meaning that they do not contain a stop codon:

1. ...A TCT AAA ATG GGT GCC...
2. ...AT CTA AAA TGG GTG CC...
3. ...ATC TAA AAT GGG TGC C...

Possible stop codons in DNA are "TGA", "TAA" and "TAG". Thus, the last reading frame in this example contains a stop codon (*TAA*), unlike the first two.

Untranslated Regions

Untranslated regions (UTRs) are sections of the mRNA before the start codon and after the stop codon that are not translated, termed the five prime untranslated region (5' UTR) and three prime untranslated region (3' UTR), respectively.

These regions are transcribed with the coding region and thus are exonic as they are present in the mature mRNA. Several roles in gene expression have been attributed to the untranslated regions, including mRNA stability, mRNA localization, and translational efficiency. The ability of a UTR to perform these functions depends on the sequence of the UTR and can differ between mRNAs.

The stability of mRNAs may be controlled by the 5' UTR and/or 3' UTR due to varying affinity for RNA degrading enzymes called ribonucleases and for ancillary proteins that can promote or inhibit RNA degradation.

Translational efficiency, including sometimes the complete inhibition of translation, can be controlled by UTRs. Proteins that bind to either the 3' or 5' UTR may affect translation by influencing the ribosome's ability to bind to the mRNA. MicroRNAs bound to the 3' UTR also may affect translational efficiency or mRNA stability.

Cytoplasmic localization of mRNA is thought to be a function of the 3' UTR. Proteins that are needed in a particular region of the cell can actually be translated there; in such a case, the 3' UTR may contain sequences that allow the transcript to be localized to this region for translation. Some of the elements contained in untranslated regions form a characteristic secondary structure when transcribed into RNA. These structural mRNA elements are involved in regulating the mRNA. Some, such as the SECIS element, are targets for proteins to bind. One class of mRNA element, the riboswitches, directly bind small molecules, changing their fold to modify levels of transcription or translation. In these cases, the mRNA regulates itself.

5' UTR

The five prime untranslated region (5' UTR), also known as the leader sequence, is a particular section of messenger RNA (mRNA) and the DNA that codes for it. At times the leader sequence can contain elements for controlling gene expression by way of attenuation. The leader sequence will be translated and the availability of the encoded amino acids will affect the transcription of the rest of the mRNA. It starts at the +1 position (where transcription begins) and ends one nucleotide before the start codon (usually AUG) of the coding region. It usually contains a ribosome binding site (RBS), in bacteria also known as the Shine Dalgarno sequence (AGGAGGU). The 5' UTR may be a hundred or more nucleotides long, and the 3' UTR may be even longer (up to several kilobases in length).

An mRNA molecule codes for a protein through translation. The mRNA also contains regions that are not translated: in eukaryotes this includes the 5' untranslated region, 3' untranslated region, 5' cap and poly-A tail.

In prokaryotic mRNA the 5' UTR is normally short. Some viruses and cellular genes have unusual long structured 5' UTRs which may have roles in gene expression.

Several regulatory sequences may be found in the 5' UTR:

- Binding sites for proteins, that may affect the mRNA's stability or translation, for example iron responsive elements, which occur in the 5' UTRs (and 3' UTRs) of a small number of eukaryotic mRNAs that regulate gene expression in response to iron.
- Regulatory elements that do not depend on proteins, such as riboswitches.
- Sequences that promote the initiation of translation.

3' UTR

The three prime untranslated region (3' UTR) is a particular section of messenger RNA (mRNA). It follows the coding region.

An mRNA molecule codes for a protein through translation. The mRNA also contains regions that are not translated. In eukaryotes these regions are the cap, 5' untranslated region, 3' untranslated region, and polyA tail. In prokaryotes mRNA structures have some differences as do histone mRNAs. However, both have 3' UTRs.

Several regulatory sequences are found in the 3' UTR:

- A polyadenylation signal, usually *AAUAAA*, or a slight variant. This marks the site of cleavage of the transcript approximately 30 base pairs past the signal, followed by the addition of several hundred adenine residues (poly-A tail).
- Binding sites for proteins, that may affect the mRNA's stability or location in the cell, like SECIS elements (which direct the ribosome to translate the codon UGA as selenocysteines rather than as a stop codon), or AU-rich elements (AREs), stretches consisting of mainly adenine and uracil nucleotides (which can either stabilize or destabilize the mRNA depending on the protein bound to it).
- Binding sites for miRNAs, a type of RNAi.

Poly(A) Tail

The 3' poly(A) tail is a long sequence of adenine nucleotides (often several hundred) added to the 3' end of the pre-mRNA. This tail promotes export from the nucleus and translation, and protects the mRNA from degradation.

Monocistronic versus Polycistronic mRNA

An mRNA molecule is said to be monocistronic when it contains the genetic information to translate only a single protein. This is the case for most of the eukaryotic mRNAs.

On the other hand, polycistronic mRNA carries the information of several genes, which are translated into several proteins. These proteins usually have a related function and are grouped and regulated together in an operon. Most of the mRNA found in bacteria and archea are polycistronic. Dicistronic or bicistronic is the term used to describe an mRNA that encodes only two proteins.

mRNA Circularization

In eukaryotes it is thought that mRNA molecules form circular structures due to an interaction between the cap binding complex and poly(A)-binding protein. Circularization is thought to promote recycling of ribosomes on the same message leading to efficient translation.

Degradation

Different mRNAs within the same cell have distinct lifetimes (stabilities). In bacterial cells, individual mRNAs can survive from seconds to more than an hour; in mammalian cells, mRNA lifetimes range from several minutes to days. The greater the stability of an mRNA, the more protein may be produced from that mRNA. The limited lifetime of mRNA enables a cell to alter protein synthesis rapidly in response to its changing needs. There are many mechanisms that lead to the destruction of a mRNA.

Prokaryotic mRNA Degradation

In prokaryotes the lifetime of mRNA is generally much shorter than in eukaryotes. Prokaryotes degrade messages by using a combination of ribonucleases, including endonucleases, 3' exonucleases, and 5' exonucleases.

In some instances, small RNA molecules (sRNA) tens to hundreds of nucleotides long can stimulate the degradation of specific mRNAs by base pairing with complementary sequences and facilitating ribonuclease cleavage. It was recently shown that bacteria also have a sort of 5' cap consisting of a triphosphate on the 5' end. Removal of two of the phosphates leaves a 5' monophosphate, causing the message to be destroyed by the endonuclease RNase E.

Eukaryotic mRNA Turnover

Inside eukaryotic cells there is a balance between the processes of translation and mRNA decay. Messages that are being actively translated are bound by ribosomes, the eukaryotic initiation factors eIF-4E and eIF-4G, and poly(A)-binding protein. eIF-4E and eIF-4G block the decapping enzyme (DCP2), and poly(A)-binding protein blocks the exosome complex, protecting the ends of the message. The balance between translation and decay is reflected in the size and abundance of cytoplasmic structures known as P-bodies The poly(A) tail of the mRNA is shortened by specialized exonucleases that are targeted to specific messenger RNAs by a combination of cis-regulatory sequences on the RNA and trans-acting RNA-binding proteins. Poly(A) tail removal is thought to disrupt the circular structure of the message and destabilize the cap binding complex. The message is then subject to degradation by either the exosome complex or the decapping complex. In this way, translationally inactive messages can be destroyed quickly, while active messages remain intact.

AU-rich Element Decay

The presence of AU-rich elements in some mammalian mRNAs tends to destabilize those transcripts through the action of cellular proteins that bind these sequences and stimulate poly(A) tail removal. Loss of the poly(A) tail is thought to promote mRNA degradation by facilitating attack by both the exosome complex and the decapping complex. Rapid mRNA degradation via AU-rich elements is a critical mechanism for preventing the overproduction of potent cytokines such as tumour necrosis factor (TNF) and granulocyte-macrophage colony stimulating factor (GM-CSF). AU-rich elements also regulate the biosynthesis of proto-oncogenic transcription factors like c-Jun and c-Fos.

Nonsense Mediated Decay

Nonsense-mediated decay (NMD) is a cellular mechanism of mRNA surveillance that functions to detect nonsense mutations and prevent the expression of truncated or erroneous proteins. Following transcription, precursor mRNA undergoes an assemblage of ribonucleoprotein (RNP) components followed by regulatory pre-mRNA processing. Large average intron size in eukaryotic cells greatly increases the probability that aberrant mRNA splicing will result in the presence of a nonsense (stop) codon (UAA, UAG, UGA) somewhere within the open reading frame. NMD is triggered by exon junction complexes (EJCs) (components of the assembled RNP) that are deposited during pre-mRNA processing. While EJCs located in an open reading frame upstream of exon-exon junctions may facilitate ribosomal recruitment prior to displacement by a "pioneer" round of translation, EJCs located downstream of a nonsense codon are not displaced because the ribosome is released from the transcript before reaching it. These remaining EJCs function as tags for recruitment of UPF1 following the mRNA's transport out of the nucleus and into the cytosol where the RNA is degraded, for example by the exosome complex. NMD is not only a mechanism for degrading aberrant mRNA, however, as there are numerous examples of normal transcripts whose expression is regulated by this process including the plasticity protein Arc/Arg3.1.

Eukaryotic messages are subject to surveillance by nonsense mediated decay (NMD), which checks for the presence of premature stop codons (nonsense codons) in the message. These can arise via incomplete splicing, V(D)J recombination in the adaptive immune system, mutations in DNA, transcription errors, leaky scanning by the ribosome causing a frame shift, and other causes. Detection of a premature stop codon triggers mRNA degradation by 5' decapping, 3' poly(A) tail removal, or endonucleolytic cleavage.

Small Interfering RNA (siRNA)

Small interfering RNA (siRNA), sometimes known as short interfering RNA or silencing RNA, is a class of double-stranded RNA molecules, 20-25 nucleotides in length, that play a variety of roles in biology. The most notable role of siRNA is its involvement in the RNA interference (RNAi) pathway, where it interferes with the expression of a specific gene. In addition to its role in the RNAi pathway, siRNA also acts in RNAi-related pathways, e.g., as an antiviral mechanism or in shaping the chromatin structure of a genome; the complexity of these pathways is only now being elucidated.

siRNAs were first discovered by David Baulcombe's group at the Sainsbury Laboratory in Norwich, England, as part of post-transcriptional gene silencing (PTGS) in plants. The group published their findings in *Science* in a paper titled "A species of small antisense RNA in posttranscriptional gene silencing in plants". Shortly thereafter, in 2001, synthetic siRNAs were shown to be able to induce RNAi in mammalian cells by Thomas Tuschl, and colleagues in a paper published in *Nature*. This discovery led to a surge in interest in harnessing RNAi for biomedical research and drug development.

RNAi Induction Using siRNAs or their Biosynthetic Precursors

Transfection of an exogenous siRNA can be problematic because the gene knockdown effect is only transient, in particular, in rapidly dividing cells. One way of overcoming this challenge is to modify the siRNA in such a way as to allow it to be expressed by an appropriate vector, e.g., a plasmid. This is done by the introduction of a loop between the two strands, thus producing a single transcript, which can be processed into a functional siRNA. Such transcription cassettes typically use an RNA polymerase III promoter (e.g., U6 or H1), which usually directs the transcription of small nuclear RNAs (snRNAs) (U6 is involved in gene splicing; H1 is the RNase component of human RNase P). It is assumed (although not known for certain) that the resulting siRNA transcript is then processed by Dicer.

RNA Activation

It has recently been found that dsRNA can also activate gene expression, a mechanism that has been termed "small RNA-induced gene activation" or RNAa. It has been shown that dsRNAs targeting gene promoters induce potent transcriptional activation of associated genes. RNAa was demonstrated in human cells using synthetic dsRNAs, termed "small activating RNAs" (saRNAs). It is currently not known whether RNAa is conserved in other organisms.

Challenges: Avoiding Nonspecific Effects

Because RNAi intersects with a number of other pathways, it is not surprising that on occasion nonspecific effects are triggered by the experimental introduction of an siRNA. When a mammalian cell encounters a double-stranded RNA such as an siRNA, it may mistake it as a viral by-product and mount an immune response. Furthermore, because structurally related microRNAs modulate gene expression largely via incomplete complementarity base pair interactions with a target mRNA, the introduction of an siRNA may cause unintended off-targeting.

Innate Immunity

Introduction of too much siRNA can result in nonspecific events due to activation of innate immune responses. Most evidence to date suggests that this is probably due to activation of the dsRNA sensor PKR, although retinoic acid-inducible gene I (RIG-I) may also be involved.

The induction of cytokines via toll-like receptor 7 (TLR7) has also been described. One promising method of reducing the nonspecific effects is to convert the siRNA into a microRNA. MicroRNAs occur naturally, and by harnessing this endogenous pathway it should be possible to achieve similar gene knockdown at comparatively low concentrations of resulting siRNAs. This should minimize nonspecific effects.

Off-targeting

Off-targeting is another challenge to the use of siRNAs as a gene knockdown tool. Here, genes with incomplete complementarity are inadvertently downregulated by the siRNA (in effect, the siRNA acts as a miRNA), leading to problems in data interpretation and potential toxicity.

This, however, can be partly addressed by designing appropriate control experiments, and siRNA design algorithms are currently being developed to produce siRNAs free from off-targeting. Genome-wide expression analysis, e.g., by microarray technology, can then be used to verify this and further refine the algorithms. A 2006 paper from the laboratory of Dr. Khvorova implicates 6- or 7-basepair-long stretches from position 2 onward in the siRNA matching with 3'UTR regions in off-targeted genes.

Possible Therapeutic Applications and Challenges

Given the ability to knock down, in essence, any gene of interest, RNAi via siRNAs has generated a great deal of interest in both basic and applied biology. There are an increasing number of large-scale RNAi screens that are designed to identify the important genes in various biological pathways. Because disease processes also depend on the activity of multiple genes, it is expected that in some situations turning off the activity of a gene with an siRNA could produce a therapeutic benefit.

However, applying RNAi via siRNAs to living animals, especially humans, poses many challenges. Under experiments, siRNAs show different effectiveness in different cell types in a manner as yet poorly understood: Some cells respond well to siRNAs and show a robust knockdown, whereas others show no such knockdown (even despite efficient transfection).

Phase I results of the first two therapeutic RNAi trials (indicated for age-related macular degeneration, aka AMD) reported at the end of 2005 that siRNAs are well tolerated and have suitable pharmacokinetic properties.

Proof of concept trials have indicated that Ebola-targeted siRNAs may be effective as post-exposure prophylaxis in humans, with 100% of non-human primates surviviing a lethal dose of Zaire Ebolavirus, the most lethal strain.

In metazoans, small interfering RNAs (siRNAs) processed by Dicer are incorporated into a complex known as the RNA-induced silencing complex or RISC. This complex contains an endonuclease that cleaves perfectly complementary messages to which the siRNA binds. The resulting mRNA fragments are then destroyed by exonucleases. siRNA is commonly used in laboratories to block the function of genes in cell culture. It is thought to be part of the innate immune system as a defense against double-stranded RNA viruses.

MicroRNA (miRNA)

MicroRNAs (miRNAs) are small RNAs that typically are partially complementary to sequences in metazoan messenger RNAs. Binding of a miRNA to a message can repress translation of that message and accelerate poly(A) tail removal, thereby hastening mRNA degradation. The mechanism of action of miRNAs is the subject of active research. MicroRNAs (miRNAs) are short ribonucleic acid (RNA) molecules, on average only 22 nucleotides long and are found in all eukaryotic cells, except fungi and marine plants. miRNAs are post-transcriptional regulators that bind to complementary sequences on target messenger RNA transcripts (mRNAs), usually resulting in translational repression and gene silencing. The human genome may encode over 1000 miRNAs, which may target about 60% of mammalian genes and are abundant in many human cell types. miRNAs show very different characteristics between plants and metazoans. In plants the miRNA complementarity to its mRNA target is nearly perfect, with no or few mismatched bases. In metazoans on the other hand miRNA complementarity is far from perfect and one miRNA can target many different sites on the same mRNA or on many different mRNAs. Another difference is the location of target sites on mRNAs. In metazoans the miRNA target sites are in the three prime untranslated regions (3'UTR) of the mRNA. In plants targets can be located in the 3' UTR but are more often in the coding region itself. MiRNAs are well conserved in eukaryotic organism and are thought to be a vital and evolutionarily ancient component of genetic regulation.

The first miRNAs were characterized in the early 1990s, but miRNAs were not recognized as a distinct class of biologic regulators with conserved functions until the early 2000s. Since then, miRNA research has revealed multiple roles in negative regulation (transcript degradation and sequestering, translational suppression) and possible involvement in positive regulation (transcriptional and translational activation). By affecting gene regulation, miRNAs are likely to be involved in most biologic processes. Different sets of expressed miRNAs are found in different cell types and tissues. Aberrant expression of miRNAs has been implicated in numerous disease states, and miRNA-based therapies are under investigation.

History

MicroRNAs were discovered in 1993 by Victor Ambros, Rosalind Lee and Rhonda Feinbaum during a study of the gene *lin-14* in *C. elegans* development. They found that LIN-14 protein abundance was regulated by a short RNA product encoded by the *lin-4* gene. A 61 nucleotide precursor from *lin-4* gene matured to a 22 nucleotide RNA containing sequences partially complementary to multiple sequences in the 3' UTR of the *lin-14* mRNA. This complementarity was sufficient and necessary to inhibit the translation of *lin-14* mRNA into LIN-14 protein. Retrospectively, the *lin-4* small RNA was the first microRNA to be identified, though at the time, it was thought to be a nematode idiosyncrasy. Only in 2000 was a second RNA characterized: let-7, which repressed *lin-41*, *lin-14*, *lin-28*, *lin-42*, and *daf-12* expression during developmental stage transitions in *C. elegans*. let-7 was soon found to be conserved in many species, indicating the existence of a wider phenomenon.

Nomenclature

Under a standard nomenclature system, names are assigned to experimentally confirmed miRNAs before publication of their discovery. The prefix "mir" is followed by a dash and a number, the latter often indicating order of naming. For example, mir-123 was named and likely discovered prior to mir-456. The uncapitalized "mir-" refers to the pre-miRNA, while a capitalized "miR-" refers to the mature form. miRNAs with nearly identical sequences bar one or two nucleotides are annotated with an additional lower case letter. For example, miR-123a would be closely related to miR-123b. Pre-miRNAs that lead to 100% identical mature miRNAs but that are located at different places in the genome are indicated with an additional dash-number suffix.

For example, the pre-miRNAs hsa-mir-194-1 and hsa-mir-194-2 lead to an identical mature miRNA (hsa-miR-194) but are located in different regions of the genome. Species of origin is designated with a three-letter prefix, e.g., hsa-miR-123 would be from human (*Homo sapiens*) and oar-miR-123 would be a sheep (*Ovis aries*) miRNA. Other common prefixes include 'v' for viral (miRNA encoded by a viral genome) and 'd' for *Drosophila* miRNA (a fruit fly commonly studied in genetic research). When two mature microRNAs originate from opposite arms of the same pre-miRNA, they are denoted with a -3p or -5p suffix. (In the past, this distinction was also made with 's' (sense) and 'as' (antisense)). When relative expression levels are known, an asterisk following the name indicates an miRNA expressed at low levels relative to the miRNA in the opposite arm of a hairpin. For example, miR-123 and miR-123* would share a pre-miRNA hairpin, but more miR-123 would be found in the cell.

Biogenesis

Most microRNA genes are found in intergenic regions or in anti-sense orientation to genes and contain their own miRNA gene promoter and regulatory units. As much as 40% of miRNA genes may lie in the introns of protein and non-protein coding genes or even in exons.

These are usually, though not exclusively, found in a sense orientation. and thus usually are regulated together with their host genes. Other miRNA genes showing a common promoter include the 42-48% of all miRNAs originating from polycistronic units containing 2-7 discrete loops from which mature miRNAs are processed, although this does not necessarily mean the mature miRNAs of a family will be homologous in structure and function. The promoters mentioned have been shown to have some similarities in their motifs to promoters of other genes transcribed by RNA polymerase II such as protein coding genes. The DNA template is not the final word on mature miRNA production: 6% of human miRNAs show RNA editing, the site-specific modification of RNA sequences to yield products different from those encoded by their DNA. This increases the diversity and scope of miRNA action beyond that implicated from the genome alone.

Transcription

miRNA genes are usually transcribed by RNA polymerase II (Pol II). The polymerase often binds to a promoter found near the DNA sequence encoding what will become the hairpin loop of the pre-miRNA. The resulting transcript is capped with a specially-modified nucleotide at the 5' end, polyadenylated with multiple adenosines (a poly(A) tail), and spliced. The product, called a primary miRNA (pri-miRNA), may be hundreds or thousands of nucleotides in length and contain one or more miRNA stem loops. When a stem loop precursor is found in the 3' UTR, a transcript may serve as a pri-miRNA and a mRNA. RNA polymerase III (Pol III) transcribes some miRNAs, especially those with upstream Alu sequences, transfer RNAs (tRNAs), and mammalian wide interspersed repeat (MWIR) promoter units.

Nuclear Processing

A single pri-miRNA may contain from one to six miRNA precursors. These hairpin loop structures are composed of about 70 nucleotides each. Each hairpin is flanked by sequences necessary for efficient processing. The double-stranded RNA structure of the hairpins in a pri-miRNA is recognized by a nuclear protein known as DiGeorge Syndrome Critical Region 8 (DGCR8 or "Pasha" in invertebrates), named for its association with DiGeorge Syndrome. DGCR8 associates with the enzyme Drosha, a protein that cuts RNA, to form the "Microprocessor" complex.

In this complex, DGCR8 orients the catalytic RNase III domain of Drosha to liberate hairpins from pri-miRNAs by cleaving RNA about eleven nucleotides from the hairpin base (two helical RNA turns into the stem). The resulting hairpin, known as a pre-miRNA (precursor-miRNA), has a two-nucleotide overhang at its 3' end; it has 3' hydroxyl and 5' phosphate groups.

Pre-miRNAs that are spliced directly out of introns, bypassing the Microprocessor complex, are known as "mirtrons." Originally thought to exist only in *Drosophila* and *C. elegans*, mirtrons have now been found in mammals.

Perhaps as many as 16% of pri-miRNAs may be altered through nuclear RNA editing. Most commonly, enzymes known as adenosine deaminases acting on RNA (ADARs) catalyze adenosine to inosine (A to I) transitions. RNA editing can halt nuclear processing (for example, of pri-miR-142, leading to degradation by the ribonuclease Tudor-SN) and alter downstream processes including cytoplasmic miRNA processing and target specificity (e.g., by changing the seed region of miR-376 in the central nervous system).

Nuclear Export

pre-miRNA hairpins are exported from the nucleus in a process involving the nucleocytoplasmic shuttle Exportin-5. This protein, a member of the *karyopherin* family, recognizes a two-nucleotide overhang left by the RNase III enzyme Drosha at the 3' end of the pre-miRNA hairpin. Exportin-5-mediated transport to the cytoplasm is energy-dependent, using GTP bound to the Ran protein.

Cytoplasmic Processing

In the cytoplasm, the pre-miRNA hairpin is cleaved by the RNase III enzyme Dicer. This endoribonuclease interacts with the 3' end of the hairpin and cuts away the loop joining the 3' and 5' arms, yielding an imperfect miRNA:miRNA* duplex about 22 nucleotides in length. Overall hairpin length and loop size influence the efficiency of Dicer processing, and the imperfect nature of the miRNA:miRNA* pairing also affects cleavage. Although either strand of the duplex may potentially act as a functional miRNA, only one strand is usually incorporated into the RNA-induced silencing complex (RISC) where the miRNA and its mRNA target interact.

Biogenesis in Plants

miRNA biogenesis in plants differs from metazoan biogenesis mainly in the steps of nuclear processing and export. Instead of being cleaved by two different enzymes, once inside and once outside the nucleus, both cleavages of the plant miRNA is performed by a Dicer homolog, called Dicer-like1 (DL1). DL1 is only expressed in the nucleus of plant cells, which indicates that both reactions take place inside the nucleus. Before plant miRNA:miRNA* duplexes are transported out of the nucleus its 3' overhangs are methylated by a RNA methyltransferaseprotein called Hua-Enhancer1 (HEN1). The duplex is then transported out of the nucleus to the cytoplasm by a protein called Hasty (HST), an Exportin 5 homolog, where they disassemble and the mature miRNA is incorporated into the RISC.

The RNA-induced Silencing Complex

The mature miRNA is part of an active RNA-induced silencing complex (RISC) containing Dicer and many associated proteins. RISC is also known as a microRNA ribonucleoprotein complex (miRNP); RISC with incorporated miRNA is sometimes referred to as "miRISC."

Dicer processing of the pre-miRNA is thought to be coupled with unwinding of the duplex. Generally, only one strand is incorporated into the miRISC, selected on the basis of its thermodynamic instability and weaker base-pairing relative to the other strand. The position of the stem-loop may also influence strand choice. The other strand, called the passenger strand due to its lower levels in the steady state, is denoted with an asterisk (*) and is normally degraded. In some cases, both strands of the duplex are viable and become functional miRNA that target different mRNA populations.

Members of the argonaute (Ago) protein family are central to RISC function. Argonautes are needed for miRNA-induced silencing and contain two conserved RNA binding domains: a PAZ domain that can bind the single stranded 3' end of the mature miRNA and a PIWI domain that structurally resembles ribonuclease-H and functions to interact with the 5' end of the guide strand. They bind the mature miRNA and orient it for interaction with a target mRNA. Some argonautes, for example human Ago2, cleave target transcripts directly; argonautes may also recruit additional proteins to achieve translational repression. The human genome encodes eight argonaute proteins divided by sequence similarities into two families: AGO (with four members present in all mammalian cells and called E1F2C/hAgo in humans), and PIWI (found in the germ line and hematopoietic stem cells).

Additional RISC components include TRBP [human immunodeficiency virus (HIV) transactivating response RNA (TAR) binding protein], PACT (protein activator of the interferon induced protein kinase (PACT), the SMN complex, fragile X mental retardation protein (FMRP), and Tudor staphylococcal nuclease-domain-containing protein (Tudor-SN).

Mode of Silencing

Gene silencing may occur either via mRNA degradation or preventing mRNA from being translated. It has been demonstrated that if there is complete complementation between the miRNA and target mRNA sequence, Ago2 can cleave the mRNA and lead to direct mRNA degradation. Yet, if there isn't complete complementation the silencing is achieved by preventing translation.

miRNA Turnover

Turnover of mature miRNA is needed for rapid changes in miRNA expression profiles. During miRNA maturation in the cytoplasm, uptake by the Argonaute protein is thought to stabilize the guide strand, while the opposite (* or "passenger") strand is preferentially destroyed. In what has been called a "Use it or lose it" strategy, Argonaute may preferentially retain miRNAs with many targets over miRNAs with few or no targets, leading to degradation of the non-targeting molecules.

Decay of mature miRNAs in animals is mediated by the 5′-to-3′ exoribonuclease XRN2, also known as Rat1p. In plants, SDN (small RNA degrading nuclease) family members degrade miRNAs in the opposite (3'-to-5') direction. Similar enzymes are encoded in animal genomes, but their roles have not yet been described.

Several miRNA modifications affect miRNA stability. As indicated by work in the model organism *Arabidopsis thaliana* (thale cress), mature plant miRNAs appear to be stabilized by the addition of methyl moieties at the 3' end. The 2'-O-conjugated methyl groups block the addition of uracil (U) residues by uridyltransferase enzymes, a modification that may be associated with miRNA degradation. However, uridylation may also protect some miRNAs; the consequences of this modification are incompletely understood. Uridylation of some animal miRNAs has also been reported. Both plant and animal miRNAs may be altered by addition of adenine (A) residues to the 3' end of the miRNA. An extra A added to the end of mammalian miR-122, a liver-enriched miRNA important in Hepatitis C, stabilizes the molecule, and plant miRNAs ending with an adenine residue have slower decay rates.

Cellular Functions

The function of miRNAs appears to be in gene regulation. For that purpose, a miRNA is complementary to a part of one or more messenger RNAs (mRNAs). Animal miRNAs are usually complementary to a site in the 3' UTR whereas plant miRNAs are usually complementary to coding regions of mRNAs. Perfect or near perfect base pairing with the target RNA promotes cleavage of the RNA.

This is the primary mode of plant microRNAs. In animals, microRNAs more often only partially base pair and inhibit protein translation of the target mRNA (this exists in plants as well but is less common). MicroRNAs that are partially complementary to a target can also speed up deadenylation, causing mRNAs to be degraded sooner.

For partially complementary microRNAs to recognise their targets, nucleotides 2–7 of the miRNA (its 'seed region') still have to be perfectly complementary. miRNAs occasionally also cause histone modification and DNA methylation of promoter sites, which affects the expression of target genes.

Animal microRNAs target in particular developmental genes. In contrast, genes involved in functions common to all cells, such as gene expression, have very few microRNA target sites and seem to be under selection to avoid targeting by microRNAs.

dsRNA can also activate gene expression, a mechanism that has been termed "small RNA-induced gene activation" or RNAa. dsRNAs targeting gene promoters can induce potent transcriptional activation of associated genes. This was demonstrated in human cells using synthetic dsRNAs termed small activating RNAs (saRNAs), but has also been demonstrated for endogenous microRNA.

Evolution

MicroRNAs are significant phylogenetic markers because of their astonishingly low rate of evolution. Their origin may have permitted the development of morphological innovation, and by making gene expression more specific and 'fine-tunable', permitted the genesis of complex organs and perhaps, ultimately, complex life. Indeed, rapid bursts of morphological innovation are generally associated with a high rate of microRNA accumulation.

MicroRNAs originate predominantly by the random formation of hairpins in "non-coding" sections of DNA (i.e. introns or intergene regions), but also by the duplication and modification of existing microRNAs. The rate of evolution (i.e. nucleotide substitution) in recently-originated microRNAs is comparable to that elsewhere in the non-coding DNA, implying evolution by neutral drift; however, older microRNAs have a much lower rate of change (often less than one substitution per hundred million years), suggesting that once a microRNA gains a function it undergoes extreme purifying selection. At this point, a microRNA is rarely lost from an animal's genome, although microRNAs which are more recently derived (and thus presumably non-functional) are frequently lost. This makes them a valuable phylogenetic marker, and they are being looked upon as a possible solution to such outstanding phylogenetic problems as the relationships of arthropods.

MicroRNAs feature in the genomes of most eukaryotic organisms, from the brown algae to the metazoa. Across all species, in excess of 5000 had been identified by March 2010. Whilst short RNA sequences (50 – hundreds of base pairs) of a broadly comparable function occur in bacteria, bacteria lack true microRNAs.

Experimental Detection and Manipulation of miRNA

MicroRNA expression can be quantified in a two-step polymerase chain reaction process of modified RT-PCR followed by quantitative real-time PCR. Variations of this method achieve absolute or relative quantification. miRNAs can also be hybridized to microarrays, slides or chips with probes to hundreds or thousands of miRNA targets, so that relative levels of miRNAs can be determined in different samples. MicroRNAs can be both discovered and profiled by high-throughput sequencing methods. The activity of an miRNA can be experimentally inhibited using a locked nucleic acid (LNA) oligo, a Morpholino oligo or a 2'-O-methyl RNA oligo. MicroRNA maturation can be inhibited at several points by steric-blocking oligos. The miRNA target site of an mRNA transcript can also be blocked by a steric-blocking oligo. Additionally, a specific miRNA can be silenced by a complementary antagomir. For the "in situ" detection of miRNA, the use of LNA is currently the only efficient method. The locked conformation of LNA results in enhanced hybridization properties and increases sensitivity and selectivity, making it ideal for detection of short miRNA.

miRNA and Disease

Just as miRNA is involved in the normal functioning of eukaryotic cells, so has dysregulation of miRNA been associated with disease. A manually curated, publicly available database miR2Disease documents known relationships between miRNA dysregulation and human disease.

miRNA and Cancer

Several miRNAs have been found to have links with some types of cancer. A study of mice altered to produce excess c-Myc — a protein with mutated forms implicated in several cancers — shows that miRNA has an effect on the development of cancer. Mice that were engineered to produce a surplus of types of miRNA found in lymphoma cells developed the disease within 50 days and died two weeks later. In contrast, mice without the surplus miRNA lived over 100 days. Leukemia can be caused by the insertion of a viral genome next to the 17-92 array of microRNAs leading to increased expression of this microRNA.

Another study found that two types of miRNA inhibit the E2F1 protein, which regulates cell proliferation. miRNA appears to bind to messenger RNA before it can be translated to proteins that switch genes on and off. By measuring activity among 217 genes encoding miRNA, patterns of gene activity that can distinguish types of cancers can be discerned. miRNA signatures may enable classification of cancer.

This will allow doctors to determine the original tissue type which spawned a cancer and to be able to target a treatment course based on the original tissue type. miRNA profiling has already been able to determine whether patients with chronic lymphocytic leukemia had slow growing or aggressive forms of the cancer. Transgenic mice that over-express or lack specific miRNAs have provided insight into the role of small RNAs in various malignancies.

A novel miRNA-profiling based screening assay for the detection of early-stage colorectal cancer has been developed and is currently in clinical trials. Early results showed that blood plasma samples collected from patients with early, resectable (Stage II) colorectal cancer could be distinguished from those of sex-and age-matched healthy volunteers. Sufficient selectivity and specificity could be achieved using small (less than 1 mL) samples of blood. The test has potential to be a cost-effective, non-invasive way to identify at-risk patients who should undergo colonoscopy.

miRNA and Heart Disease

The global role of miRNA function in the heart has been addressed by conditionally inhibiting miRNA maturation in the murine heart, and has revealed that miRNAs play an essential role during its development. miRNA expression profiling studies demonstrate that expression levels of specific miRNAs change in diseased human hearts, pointing to their involvement in cardiomyopathies.

Furthermore, studies on specific miRNAs in animal models have identified distinct roles for miRNAs both during heart development and under pathological conditions, including the regulation of key factors important for cardiogenesis, the hypertrophic growth response, and cardiac conductance.

miRNA and the Nervous System

miRNAs appear to regulate the nervous system. Neural miRNAs are involved at various stages of synaptic development, including dendritogenesis (involving miR-132, miR-134 and miR-124), synapse formation and synapse maturation (where miR-134 and miR-138 are thought to be involved). Some studies find altered miRNA expression in schizophrenia.

miRNA and Non-coding RNAs

When the human genome project mapped its first chromosome in 1999, it was predicted the genome would contain over 100,000 protein coding genes. However, only around 20,000 were eventually identified (International Human Genome Sequencing Consortium, 2004).

Since then, the advent of bioinformatics approaches combined with genome tiling studies examining the transcriptome, systematic sequencing of full length cDNA libraries, and experimental validation (including the creation of miRNA derived antisense oligonucleotides called antagomirs) have revealed that many transcripts are non protein-coding RNA, including several snoRNAs and miRNAs.

Other Decay Mechanisms

There are other ways in which messages can be degraded, including non-stop decay, silencing by Piwi-interacting RNA (piRNA), and surely other means.

6

Clinical Trials

On January 23, 2009, the US Food and Drug Administration gave clearance to Geron Corporation for the initiation of the first clinical trial of an embryonic stem cell-based therapy on humans. The trial will evaluate the drug GRNOPC1, embryonic stem cell-derived oligodendrocyte progenitor cells, on patients with acute spinal cord injury. As of mid 2010 hundreds of phase III clinical trials involving stem cells have been registered.

Stem Cell Use in Animals

Veterinary Applications Potential Contributions to Veterinary Medicine

Research currently conducted on horses, dogs, and cats can benefit the development of stem-cell treatments in veterinary medicine and can target a wide range of injuries and diseases such as myocardial infarction, stroke, tendon and ligament damage, osteoarthritis, osteochondrosis and muscular dystrophy both in large animals, as well as humans. While investigation of cell-based therapeutics generally reflects human medical needs, the high degree of frequency and severity of certain injuries in racehorses has put veterinary medicine at the forefront of this novel regenerative approach. Companion animals can serve as clinically relevant models that closely mimic human disease.

Development of Regenerative Treatment Models

Veterinary applications of stem cell therapy as a means of tissue regeneration have been largely shaped by research that began with the use of adult-derived mesenchymal stem cells to treat animals with injuries or defects affecting bone, cartilage, ligaments and/or tendons. Because mesenchymal stem cells can differentiate into the cells that make up bone, cartilage, tendons, and ligaments (as well as muscle, fat, and possibly other tissues), they have been the main type of stem cells studied in the treatment of diseases affecting these tissues. Mesenchymal stem cells are primarily derived from adipose tissue or bone marrow. Since an elevated immune response following cell transplantation may result in rejection of exogenous cells (except in the case of cells derived from a very closely genetically related individual), mesenchymal stem cells are often derived from the patient prior to injection in a process known as autologous transplantation. Surgical repair of bone fractures in dogs and sheep has demonstrated that engraftment of mesenchymal stem cells derived from a genetically different donor within the same species, termed allogeneic transplantation, does not elicit an immunological response in the recipient animal and can mediate regeneration of bone tissue in major bony fractures and defects.

Stem cells can speed up bone repair in fractures/defects that would normally require extensive grafting, suggesting that mesenchymal stem cell use may provide a useful alternative to conventional grafting techniques. Treating tendon and ligament injuries in horses using stem cells, whether derived from adipose tissue or bone-marrow, has support in the veterinary literature. While further studies are necessary to fully characterize the use of cell-based therapeutics for treatment of bone fractures, stem cells are thought to mediate repair via five primary mechanisms: 1) providing an antiinflammatory effect, 2) homing to damaged tissues and recruiting other cells, such as endothelial progenitor cells, that are necessary for tissue growth, 3) supporting tissue remodeling over scar formation, 4) inhibiting apoptosis, and 5) differentiating into bone, cartilage, tendon, and ligament tissue.

Significance of Stem Cell Microenvironments

The microenvironment into which stem cells are transplanted significantly alters the capacity of engrafted cells for recovery and repair. The microenviroment provides growth factors and other chemical signals that guide appropriate differentiation of transplanted cell populations and direct transplanted cells to sites of trauma or disease. Repair and recovery can then be mediated via three primary mechanisms: 1) formation and/or recruitment of new blood cells to the damaged region; 2) prevention of programed cell death or apoptosis; and 3) suppression of inflammation. To further enrich blood supply to the damaged areas, and consequently promote tissue regeneration, platelet-rich plasma could be used in conjunction with stem cell transplantation. The efficacy of some stem cell populations may also be affected by the method of delivery; for instance, to regenerate bone, stem cells are often introduced in a scaffold where they produce the minerals necessary for generation of functional bone.

Sources of Autologous (Patient-derived) Stem Cells

Autologous stem cells intended for regenerative therapy are generally isolated either from the patient's bone marrow or from adipose tissue. The number of stem cells transplanted into damaged tissue may alter efficacy of treatment.

Accordingly, stem cells derived from bone marrow aspirates, for instance, are cultured in specialized laboratories for expansion to millions of cells. Although adipose-derived tissue also requires processing prior to use, the culturing methodology for adipose-derived stem cells is not as extensive as that for bone marrow-derived cells. While it is thought that bone-marrow derived stem cells are preferred for bone, cartilage, ligament, and tendon repair, others believe that the less challenging collection techniques and the multi-cellular microenvironment already present in adipose-derived stem cell fractions make the latter the preferred source for autologous transplantation.

Currently available treatments for horses and dogs suffering from orthopaedic conditions Autologous or allogeneic stem cells are currently used as an adjunctive therapy in the surgical repair of some types of fractures in dogs and horses. Autologous stem cell-based treatments for ligament injury, tendon injury, osteoarthritis, osteochondrosis, and sub-chondral bone cysts have been commercially available to practicing veterinarians to treat horses since 2003 in the United States and since 2006 in the United Kingdom. Autologous stem-cell based treatments for tendon injury, ligament injury, and osteoarthritis in dogs have been available to veterinarians in the United States since 2005. Over 3000 privately-owned horses and dogs have been treated with autologous adipose-derived stem cells. The efficacy of these treatments has been shown in double-blind clinical trials for dogs with osteoarthritis of the hip and elbow and horses with tendon damage. The efficacy of using stem cells, whether adipose-derived or bone-marrow derived, for treating tendon and ligament injuries in horses has support in the veterinary literature.

Developments in Stem Cell Treatments in Veterinary Internal Medicine Currently, research is being conducted to develop stem cell treatments for: 1) horses suffering from COPD, neurologic disease, and laminitis; and 2) dogs and cats suffering from heart disease, liver disease, kidney disease, neurologic disease, and immune-mediated disorders.

Embryonic Stem Cell Controversy

There is widespread controversy over the use of human embryonic stem cells. This controversy primarily targets the techniques used to derive new embryonic stem cell lines, which often requires the destruction of the blastocyst. Opposition to the use of human embryonic stem cells in research is often based on philosophical, moral or religious objections.

Stem Cell Treatments Around the World

China

Stem cell research and treatment is currently being practiced at a clinical level in the People's Republic of China. The Ministry of Health of the People's Republic of China has permitted the use of stem cell therapy for conditions beyond those approved of in Western countries such as the United States, United Kingdom, and Australia. The Western World has scrutinized China for its failed attempts to meet international documentation standards of these trials and procedures, despite the overwhelmingly positive anecdotal results.

Stem cell therapies provided in China utilize a variety of cell types including umbilical cord stem cells and olfactory ensheathing cells. The stem cells are then expanded in centralized blood banks before being used in stem cell treatments. State-funded companies based in the Shenzhen Hi-Tech Industrial Zone treat the symptoms of numerous disorders with adult stem cell therapy. Hospitals throughout eastern China provide numerous therapies to patients in coordination with the stem cell providers.

These companies' therapies are currently focused on the treatment of neurodegenerative and cardiovascular disorders. The most radical successes of Chinese adult stem cell therapy have been in treating the brain. These therapies administer stem cells directly to the brain to promote greater motor and brain function in patients with Cerebral Palsy, Alzheimer's, and brain injuries.

However, retrospective studies have shown that Chinese use of fetal-derived brain tissue in spinal cord injured human subjects were not as promising as once thought: the phenotype and the fate of the transplanted cells, described as olfactory ensheathing cells, were unknown. As well, perioperative morbidity and lack of functional benefit were identified as the most serious clinical shortcomings. Furthermore, the extent of regulatory policy in the use of stem cell therapies in China is unclear. In the absence of a valid clinical trials protocol, and more regulatory oversight, Western regulatory agencies advise patients and physicians to be cautious when selecting Chinese stem cell therapeutic centers.

Mexico

Stem cell treatment is currently being practiced at a clinical level in Mexico. An International Health Department Permit (COFEPRIS) is required.

This permit allows the use of stem cell types beyond those approved of in Western countries such as the United States or Europe. Stem cell therapies provided in Mexico utilize patient Adipose, Bone Marrow, or Donor Placenta sources.

South Korea

In 2005, South Korean scientists claimed to have generated stem cells that were tailored to match the recipient. Each of the 11 new stem cell lines was developed using somatic cell nuclear transfer (SCNT) technology. The resultant cells were thought to match the genetic material of the recipient, thus suggesting minimal to no cell rejection.

This study, however, was eventually discredited as the primary researcher,Dr. Woo Suk Hwang, admitted to using cells obtained from his research staff. In Dec 2005, claims were put forward that his research had been manipulated to wrongfully indicate positive results. Later that month, these claims were confirmed by an academic panel.

Ukraine

Today, Ukraine is permitted to perform clinical trials of stem cell treatments (Order of the MH of Ukraine [1] 630 "About carrying out clinical trials of stem cells", 2008) for the treatment of these pathologies: pancreatic necrosis, cirrhosis, hepatitis, burn disease, diabetes, multiple sclerosis, critical lower limb ischemia. The first medical institution granted the right to conduct clinical trials became the "Institute of Cell Therapy"(Kiev).

- Cardiovascular Cell Therapy Research Network (CCTRN)
- Fetal tissue implant
- Human Stem Cells Institute

Pharmacological

As of August 2008, several trials of various compounds are in development or ongoing, a few at the point of testing on larger numbers of people, known as phase III of clinical trials. Substances that have shown promise in initial experiments include dopamine receptor blockers, select dopamine antagonists, such as creatine and CoQ10, the antibiotic Minocycline, antioxidant-containing foods and nutrients, and antidepressants, including selective serotonin reuptake inhibitors such as sertraline, fluoxetine, and paroxetine. In 2010, minocycline was found to be ineffective for humans in a multi-center trial.

A 2009 study suggested strong association between specific single nucleotide polymorphisms alleles and CAG expansion also provides an opportunity of personalized therapeutics in HD where the clinical development of only a small number of allele-specific targets may be sufficient to treat up to 88% of the HD patient population.

Other

Recently it was shown that increasing BDNF expression may represent a useful strategy for Huntington's disease treatment.

- Gauchers disease is a genetic disease affecting metabolism. It is more treatable then most other genetic diseases, and can be treated with enzyme replacement therapy, medication miglustat, and bone marrow transplantion.

Classification

Gaucher's disease has three common clinical subtypes.

- Type I (or non-neuropathic type) is the most common form of the disease, occurring in approximately 1 in 50,000 live births. It occurs most often among persons of Ashkenazi Jewish heritage. Symptoms may begin early in life or in adulthood and include enlarged liver and grossly enlarged spleen (together hepatosplenomegaly); the spleen can rupture and cause additional complications. Skeletal weakness and bone disease may be extensive. Spleen enlargement and bone marrow replacement cause anemia, thrombocytopenia and leukopenia. The brain is not affected pathologically, but there may be lung and, rarely, kidney impairment. Patients in this group usually bruise easily (due to low levels of platelets) and experience fatigue due to low numbers of red blood cells. Depending on disease onset and severity, type 1 patients may live well into adulthood. Many patients have a mild form of the disease or may not show any symptoms.
- Type II (or acute infantile neuropathic Gaucher's disease) typically begins within 6 months of birth and has an incidence rate of approximately 1 in 100,000 live births. Symptoms include an enlarged liver and spleen, extensive and progressive brain damage, eye movement disorders, spasticity, seizures, limb rigidity, and a poor ability to suck and swallow. Affected children usually die by age 2.
- Type III (the chronic neuropathic form) can begin at any time in childhood or even in adulthood, and occurs in approximately 1 in 100,000 live births. It is characterized by slowly progressive but milder neurologic symptoms compared to the acute or type 2 version. Major symptoms include an enlarged spleen and/or liver, seizures, poor coordination, skeletal irregularities, eye movement disorders, blood disorders including anemia and respiratory problems. Patients often live into their early teen years and adulthood.

These subtypes have come under some criticism for not taking account of the full spectrum of observable symptoms (the phenotypes.) There are also compound heterozygous variations which considerably increase the complexity of predicting disease course.

Signs and Symptoms

- Painless hepatomegaly and splenomegaly: the size of the spleen can be 1500-3000 ml, as opposed to the normal size of 50-200 ml. Splenomegaly may decrease the affected individual's capacity for eating by exerting pressure on the stomach. While painless, enlargement of spleen increases the risk of splenic rupture.

- Hypersplenism and pancytopenia: the rapid and premature destruction of blood cells, leading to anemia, neutropenia, leukopenia, and thrombocytopenia (with an increased risk of infection and bleeding).
- Cirrhosis of the liver is rare
- Severe pain associated with joints and bones, frequently presenting in hips and knees.
- Neurological symptoms occur only in some types of Gaucher's:
  - o Type II: serious convulsions, hypertonia, mental retardation, apnea.
  - o Type III: muscle twitches known as myoclonus, convulsions, dementia, ocular muscle apraxia.
- Osteoporosis: 75% develop visible bony abnormalities due to the accumulated glucosylceramide. A deformity of the distal femur in the shape of an Erlenmeyer flask is commonly described (aseptic necrosis of the femur joint).
- Yellowish-brown skin pigmentation

Pathophysiology

The disease is caused by a defect in housekeeping gene lysosomal glucocerebrosidase (also known as beta-glucosidase, EC 3.2.1.45, PDB 1OGS) on the first chromosome (1q21). The enzyme is a 55.6 KD, 497 amino acids long protein that catalyses the breakdown of glucosylceramidase, a cell membrane constituent of red and white blood cells.

The macrophages that clear these cells are unable to eliminate the waste product, which accumulates in fibrils, and turn into Gaucher cells, which appear on light microscopy to resemble crumpled-up paper.

In the brain (type II and III), glucosylceramidase accumulates due to the turnover of complex lipids during brain development and the formation of the myelin sheath of nerves.

Different mutations in the beta-glucosidase determine the remaining activity of the enzyme, and, to a large extent, the phenotype.

Heterozygotes for particular acid beta-glucosidase mutations carry about a fivefold risk of developing Parkinson's disease, making this the most common known genetic risk-factor for Parkinson's. A study of 1525 Gaucher patients in the United States suggested that while cancer risk is not elevated, particular malignancies (non-Hodgkin lymphoma, melanoma and pancreatic cancer) occurred at a 2-3 times higher rate.

Genetics

The three types of Gaucher's disease are inherited in an autosomal recessive fashion. Both parents must be carriers in order for a child to be affected. If both parents are carriers, there is a one in four, or 25%, chance with each pregnancy for an affected child. Genetic counseling and genetic testing is recommended for families who may be carriers of mutations.

Each type has been linked to particular mutations. In all, there are about 80 known mutations, grouped into three main types:

- Type I (N370S homozygote), the most common, also called the "non-neuropathic" type occurs mainly in Ashkenazi Jews, at 100 times the occurrence in the general populace. The median age at diagnosis is 28 years of age, and life expectancy is mildly decreased. There are no neurological symptoms.
- Type II (1 or 2 alleles L444P) is characterized by neurological problems in small children. The enzyme is hardly released into the lysosomes. Prognosis is dismal: most die before reaching the third birthday.
- Type III (also 1-2 copies of L444P, possibly delayed by protective polymorphisms) occurs in Swedish patients from the Norrbotten region. This group develops the disease somewhat later, but most die before their 30th birthday.

Diaz et al. suggest that the Gaucher-causing mutations entered the Ashkenazi Jewish gene pool in the early Middle Ages (48-55 generations ago).

Diagnosis

Micrograph showing crinkled paper macrophages in the marrow space in a case of Gaucher disease. H&E stain.

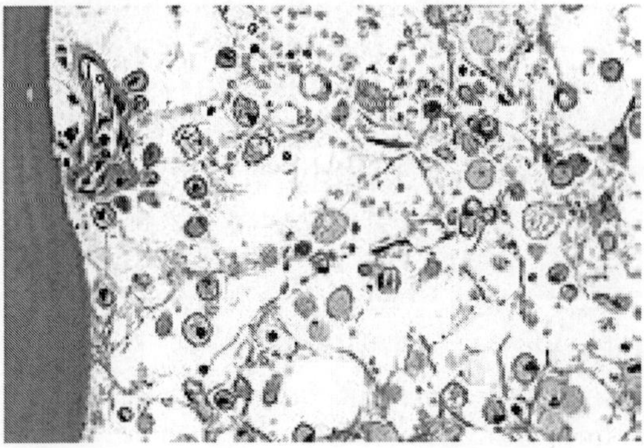

A definitive diagnosis is made with genetic testing. As there are numerous different mutations, sequencing of the beta-glucosidase gene is sometimes necessary to confirm the diagnosis. Prenatal diagnosis is available, and is useful when there is a known genetic risk factor. A diagnosis can also be implied by biochemical abnormalities such as high alkaline phosphatase, angiotensin-converting enzyme (ACE) and immunoglobulin levels, or by cell analysis showing "crinkled paper" cytoplasm and glycolipid-laden macrophages. Some lysosomal enzymes are elevated, including tartrate-resistant acid phosphatase, hexosaminidase, and a human chitinase, chitotriosidase. This latter enzyme has proved to be very useful for monitoring Gaucher's disease activity in response to treatment, and may reflect the severity of the disease

Treatment

For type 1 and most type 3 patients, enzyme replacement treatment with intravenous recombinant glucocerebrosidase (imiglucerase) can dramatically decrease liver and spleen size, reduce skeletal abnormalities, and reverse other manifestations. This treatment costs approximately $200,000 annually for a single patient and should be continued for life.

The rarity of the disease means that dose-finding studies have been difficult to conduct, so there remains controversy over the optimal dose and dosing frequency. Due to the low incidence, this has become an orphan drug in many countries, meaning that a government recognizes and accommodates the financial constraints that limit research into drugs that address a small population.Velaglucerase alfa was approved by the Food and Drug Administration (FDA) as an alternative treatment on February 26, 2010.

Successful bone marrow transplantation cures the non-neurological manifestations of the disease, because it introduces a monocyte population with active beta-glucosidase. However, this procedure carries significant risk and is rarely performed in Gaucher patients. Surgery to remove the spleen (splenectomy) may be required on rare occasions if the patient is anemic or when the enlarged organ affects the patient's comfort.

Blood transfusion may benefit some anemic patients. Other patients may require joint replacement surgery to improve mobility and quality of life. Other treatment options include antibiotics for infections, antiepileptics for seizures, bisphosphonates for bone lesions, and liver transplants. Substrate reduction therapy may prove to be effective in stopping Type 2, as it can cross through the blood barrier into the brain. There is currently no effective treatment for the severe brain damage that may occur in patients with types 2 and 3 Gaucher disease. Gene therapy may be a future step.

Gaucher's disease has recently become a target for more than one effort at pharmacological chaperoning, which involves the use of orally administered drugs that operate at a molecular level. Miglustat is one of these oral drugs. It was approved for the treatment of this disease in 2003. As of June 2009, another oral drug, isofagomine tartrate, is under development.

Epidemiology

The National Gaucher Foundation states that around 1 in 100 people in the general U.S. population is a carrier for type 1 Gaucher's disease, giving a prevalence of 1 in 40,000: among Ashkenazi Jews the rate of carriers is considerably higher, at roughly 1 in 15.

Type 2 Gaucher's disease shows no particular preference for any ethnic group.

Type 3 Gaucher's disease is especially common in the population of the Northern Swedish region of Norrbotten where the incidence of the disease is 1 in 50,000.

History

The disease was first recognised by the French doctor Philippe Gaucher, who originally described it in 1882 and lent his name to the condition. The biochemical basis for the disease would be elucidated in 1965. The first effective treatment for the disease, the drug Ceredase, was approved by the FDA in April 1991. An improved drug, Cerezyme, was approved by the FDA in May 1994 and has replaced the use of Ceredase.

Genetic epidemiology

Genetic epidemiology is the study of the role of genetic factors in determining health and disease in families and in populations, and the interplay of such genetic factors with environmental factors. In slightly more formal language, genetic epidemiology was defined by Morton as "a science which deals with the etiology, distribution, and control of disease in groups of relatives and with inherited causes of disease in populations". It is closely allied to both molecular epidemiology and statistical genetics, but these overlapping fields each have distinct emphases, societies and journals.

Traditionally, the study of the role of genetics in disease progresses through the following study designs, each answering a slightly different question:

- Familial aggregation studies: Is there a genetic component to the disease, and what are the relative contributions of genes and environment?
- Segregation studies: What is the pattern of inheritance of the disease (e.g. dominant or recessive)?
- Linkage studies: On which part of which chromosome is the disease gene located?
- Association studies: Which allele of which gene is associated with the disease?

This traditional approach has proved highly successful in identifying monogenic disorders and locating the genes responsible.

More recently, the scope of genetic epidemiology has expanded to include common diseases for which many genes each make a smaller contribution (polygenic, multifactorial or multigenic disorders). This has developed rapidly in the first decade of the 21st century following completion of the Human Genome Project, as advances in genotyping technology and associated reductions in cost has made it feasible to conduct large-scale genome-wide association studies that genotype many thousands of single nucleotide polymorphisms in thousands of individuals. These have led to the discovery of many genetic polymorphisms that influence the risk of developing many common diseases.

Genotoxicity Bioindicator

Cluster analysis in epidemiology is the method for identifying wide spread population damage (as mercury contamination in Minamata Bay) but a genotoxicity bioindicator is a preventive way of seeing congenital malformations epidemic in human populations.

A biological monitor, or biomonitor, is defined as an organism that provides quantitative information on the quality of the environment around it. Therefore, a good biomonitor will indicate the presence of the pollutant and also attempt to provide additional information about the amount and intensity of the exposure.

Endocrine Disruption (feminization); Reproductive Failure causes include early embryonic death, embryo toxicity and abnormal parental behaviour during incubation; Eggshell Thinning including Double-crested Cormorants, Ospreys, Bald Eagles, Black-crowned Night-Herons and Herring Gulls, maybe are caused by pollution or another genotoxic factor. Malformed frogs are caused by parasites, genetic disease or pollution? Crossed bills, jaw defects, extra limbs, and malformed feet, joints and eyes were found in Herring Gulls and at least eight other species of fish-eating waterbirds from great lakes, malformed frogs were found in Minnesota.

Inborn errors of metabolism

Inborn errors of metabolism comprise a large class of genetic diseases involving disorders of metabolism. The majority are due to defects of single genes that code for enzymes that facilitate conversion of various substances (substrates) into others (products). In most of the disorders, problems arise due to accumulation of substances which are toxic or interfere with normal function, or to the effects of reduced ability to synthesize essential compounds. Inborn errors of metabolism are now often referred to as congenital metabolic diseases or inherited metabolic diseases.

The term inborn error of metabolism was coined by a British physician, Archibald Garrod (1857–1936), in the early 20th century (1908). He is known for work that prefigured the "one gene-one enzyme" hypothesis, based on his studies on the nature and inheritance of alkaptonuria. His seminal text, Inborn Errors of Metabolism was published in 1923.

Major Categories of Inherited Metabolic Diseases

Traditionally the inherited metabolic diseases were categorized as disorders of carbohydrate metabolism, amino acid metabolism, organic acid metabolism, or lysosomal storage diseases. In recent decades, hundreds of new inherited disorders of metabolism have been discovered and the categories have proliferated. Following are some of the major classes of congenital metabolic diseases, with prominent examples of each class.

Many others do not fall into these categories. ICD-10 codes are provided where available:

- Disorders of carbohydrate metabolism
  - o E.g., glycogen storage disease
- Disorders of amino acid metabolism
  - o E.g., phenylketonuria, maple syrup urine disease, glutaric acidemia type 1
- Disorders of organic acid metabolism (organic acidurias)
  - o E.g., alcaptonuria
- Disorders of fatty acid oxidation and mitochondrial metabolism
  - o E.g., Medium-chain acyl-coenzyme A dehydrogenase deficiency (often shortened to MCADD.)
- Disorders of porphyrin metabolism
  - o E.g., acute intermittent porphyria
- Disorders of purine or pyrimidine metabolism
  - o E.g., Lesch-Nyhan syndrome
- Disorders of steroid metabolism
  - o E.g., congenital adrenal hyperplasia
- Disorders of mitochondrial function
  - o E.g ., Kearns-Sayre syndrome
- Disorders of peroxisomal function
  - o E.g., Zellweger syndrome
- Lysosomal storage disorders
  - o E.g., Gaucher's disease
  - o E.g., Niemann Pick disease

Incidence

In a study in British Columbia, the overall incidence of the inborn errors of metabolism were estimated to be 70 per 100,000 live births or 1 in 1,400 births, overall representing more than approximately 15% of single gene disorders in the population.

Type of Inborn Error	*Incidence*	
Disease involving amino acids (e.g. PKU), organic acids, primary lactic acidosis, galactosemia, or a urea cycle disease	24 per 100 000 births	1 in 4,200
Lysosomal storage disease	8 per 100 000 births	1 in 12,500
Peroxisomal disorder	~3 to 4 per 100 000 of births	~1 in 30,000
Respiratory chain-based mitochondrial disease	~3 per 100 000 births	1 in 33,000
Glycogen storage disease	2.3 per 100 000 births	1 in 43,000

Manifestations and Presentations

Because of the enormous number of these diseases and wide range of systems affected, nearly every "presenting complaint" to a doctor may have a congenital metabolic disease as a possible cause, especially in childhood. The following are examples of potential manifestations affecting each of the major organ systems: many manifestations may develop

- Growth failure, failure to thrive, weight loss
- Ambiguous genitalia, delayed puberty, precocious puberty
- Developmental delay, seizures, dementia, encephalopathy, stroke
- Deafness, blindness, pain agnosia
- Skin rash, abnormal pigmentation, lack of pigmentation, excessive hair growth, lumps and bumps
- Dental abnormalities
- Immunodeficiency, thrombocytopenia, anemia, enlarged spleen, enlarged lymph nodes
- Many forms of cancer
- Recurrent vomiting, diarrhea, abdominal pain
- Excessive urination, renal failure, dehydration, edema
- Hypotension, heart failure, enlarged heart, hypertension, myocardial infarction

- Hepatomegaly, jaundice, liver failure
- Unusual facial features, congenital malformations
- Excessive breathing (hyperventilation), respiratory failure
- Abnormal behavior, depression, psychosis
- Joint pain, muscle weakness, cramps
- Hypothyroidism, adrenal insufficiency, hypogonadism, diabetes mellitus

Diagnostic Techniques

Dozens of congenital metabolic diseases are now detectable by newborn screening tests, especially the expanded testing using mass spectrometry.

This is an increasingly common way for the diagnosis to be made and sometimes results in earlier treatment and a better outcome. There is a revolutionary GC/MS based technology with an integrated analytics system, which has now made it possible to test a newborn for over 100 genetic metabolic disorders.

Because of the multiplicity of conditions, many different diagnostic tests are used for screening. An abnormal result is often followed by a subsequent "definitive test" to confirm the suspected diagnosis.

Common screening tests used in the last sixty years:

- Ferric chloride test (turned colors in reaction to various abnormal metabolites in urine)
- Ninhydrin paper chromatography (detected abnormal amino acid patterns)
- Guthrie bacterial inhibition assay (detected a few amino acids in excessive amounts in blood) The dried blood spot can be used for multianalyte testing using Tandem Mass Spectrometry (MS/MS). This given an indication for a disorder. The same has to be further confirmed by enzyme assays, GC/MS or DNA Testing.
- Quantitative measurement of amino acids in plasma and urine
- Urine organic acid analysis by Gas chromatography-mass spectrometry
- Plasma acylcarnitines analysis by mass spectrometry
- Urine purines and pyrimidines analysis by Gas chromatography-mass spectrometry

Specific diagnostic tests (or focused screening for a small set of disorders):

- Tissue biopsy or necropsy: liver, muscle, brain, bone marrow
- Skin biopsy and fibroblast cultivation for specific enzyme testing
- Specific DNA testing

Treatment

In the middle of the 20th century the principal treatment for some of the amino acid disorders was restriction of dietary protein and all other care was simply management of complications. In the past twenty years, enzyme replacement, gene transfer, and organ transplantation have become available and beneficial for many previously untreatable disorders. Some of the more common or promising therapies are listed:

- Dietary restriction
  - E.g., reduction of dietary protein remains a mainstay of treatment for phenylketonuria and other amino acid disorders
- Dietary supplementation or replacement
  - E.g., oral ingestion of cornstarch several times a day helps prevent people with glycogen storage diseases from becoming seriously hypoglycemic.
- Vitamins
  - E.g., thiamine supplementation benefits several types of disorders that cause lactic acidosis.
- Intermediary metabolites, compounds, or drugs that facilitate or retard specific metabolic pathways
- Dialysis
- Enzyme replacement E.g. Acid-alpha glucosidase for Pompe disease
- Gene transfer
- Bone marrow or organ transplantation
- Treatment of symptoms and complications
- Prenatal diagnosis and avoidance of pregnancy or abortion of an affected fetus

Resources

For clinicians and scientists in the field of inborn errors of metabolism, good resources include books by Scriver, Fernandes, Clarke, Blau (diagnosis), Blau (treatment), Lyon, Nyhan, Hoffmann, and Zschocke. Other resources include genetests, orphanet, OMIM, Metab-L, societies such as the SSIEM, the SIMD and links therein. For medical students and clinicians looking for overviews of the field, such reviews can be found on pubmed and in good pediatric textbooks (e.g. articles by Saudubray, Ellaway, Raghuveer or Burton, and textbooks by Hay or Behrman).

For patients, their families or other individuals seeking good information and support groups, the National Institutes of Health offers the office of rare diseases, genetics home reference, medlineplus and health information. The National Human Genome Research Institute hosts an information center, a section for patients and the public and additional educational resources. Support groups can be found at NORD, Genetic Alliance and Orphanet. The genetic education center at the KUMC has many more useful links.

Medical Genetics

Medical genetics is the specialty of medicine that involves the diagnosis and management of hereditary disorders. Medical genetics differs from Human genetics in that human genetics is a field of scientific research that may or may not apply to medicine, but medical genetics refers to the application of genetics to medical care.

For example, research on the causes and inheritance of genetic disorders would be considered within both human genetics and medical genetics, while the diagnosis, management, and counseling of individuals with genetic disorders would be considered part of medical genetics.

In contrast, the study of typically non-medical phenotypes such as the genetics of eye color would be considered part of human genetics, but not necessarily relevant to medical genetics (except in situations such as albinism). Genetic medicine is a newer term for medical genetics and incorporates areas such as gene therapy, personalized medicine, and the rapidly emerging new medical specialty, predictive medicine.

Scope

Medical genetics encompasses many different areas, including clinical practice of physicians, genetic counselors, and nutritionists, clinical diagnostic laboratory activities, and research into the causes and inheritance of genetic disorders. Examples of conditions that fall within the scope of medical genetics include birth defects and dysmorphology, mental retardation, autism, and mitochondrial disorders, skeletal dysplasia, connective tissue disorders, cancer genetics, teratogens, and prenatal diagnosis.

Medical genetics is increasingly becoming relevant to many common diseases. Overlaps with other medical specialties are beginning to emerge, as recent advances in genetics are revealing etiologies for neurologic, cndocrine, cardiovascular, pulmonary, ophthalmologic, renal, psychiatric, and dermatologic conditions.

Subspecialties

In some ways, many of the individual fields within medical genetics are hybrids between clinical care and research. This is due in part to recent advances in science and technology (for example, the Human genome project) that have enabled an unprecedented understanding of genetic disorders.

Clinical Genetics

Clinical genetics is the practice of clinical medicine with particular attention to hereditary disorders. Referrals are made to genetics clinics for a variety of reasons, including birth defects, developmental delay, autism, epilepsy, short stature, and many others. Examples of genetic syndromes that are commonly seen in the genetics clinic include chromosomal rearrangements, Down syndrome, DiGeorge syndrome (22q11.2 Deletion Syndrome), Fragile X syndrome, Marfan syndrome, Neurofibromatosis, Turner syndrome, and Williams syndrome.

Metabolic/Biochemical Genetics

Metabolic (or biochemical) genetics involves the diagnosis and management of inborn errors of metabolism in which patients have enzymatic deficiencies that perturb biochemical pathways involved in metabolism of carbohydrates, amino acids, and lipids. Examples of metabolic disorders include galactosemia, glycogen storage disease, lysosomal storage disorders, metabolic acidosis, peroxisomal disorders, phenylketonuria, and urea cycle disorders.

Cytogenetics

Cytogenetics is the study of chromosomes and chromosome abnormalities. While cytogenetics historically relied on microscopy to analyze chromosomes, new molecular technologies such as array comparative genomic hybridization are now becoming widely used. Examples of chromosome abnormalities include aneuploidy, chromosomal rearrangements, and genomic deletion/duplication disorders.

Molecular Genetics

Molecular genetics involves the discovery of and laboratory testing for DNA mutations that underlie many single gene disorders. Examples of single gene disorders include achondroplasia, cystic fibrosis, Duchenne muscular dystrophy, hereditary breast cancer (BRCA1/2), Huntington disease, Marfan syndrome, Noonan syndrome, and Rett syndrome. Molecular tests are also used in the diagnosis of syndromes involving epigenetic abnormalities, such as Angelman syndrome, Beckwith-Wiedemann syndrome, Prader-willi syndrome, and uniparental disomy.

Mitochondrial genetics

Mitochondrial genetics concerns the diagnosis and management of mitochondrial disorders, which have a molecular basis but often result in biochemical abnormalities due to deficient energy production. There exists some overlap between medical genetic diagnostic laboratories and molecular pathology.

Genetic Counseling

Genetic counseling is the process of providing information about genetic conditions, diagnostic testing, and risks in other family members, within the framework of nondirective counseling. Genetic counselors are non-physician members of the medical genetics team who specialize in family risk assessment and counseling of patients regarding genetic disorders. The precise role of the genetic counselor varies somewhat depending on the disorder.

History

Although genetics has its roots back in the 19th century with the work of the Bohemian monk Gregor Mendel and other pioneering scientists, human genetics emerged later. It started to develop, albeit slowly, during the first half of the 20th century. Mendelian (single-gene) inheritance was studied in a number of important disorders such as albinism, brachydactyly (short fingers and toes), and hemophilia. Mathematical approaches were also devised and applied to human genetics. Population genetics was created.

Medical genetics was a late developer, emerging largely after the close of World War II (1945) when the eugenics movement had fallen into disrepute. The Nazi misuse of eugenics sounded its death knell. Shorn of eugenics, a scientific approach could be used and was applied to human and medical genetics. Medical genetics saw an increasingly rapid rise in the second half of the 20th century and continues in the 21st century.

Notable Practitioners

- Victor McKusick (1921–2008), "The Father of Genetic Medicine" and principal author of Mendelian Inheritance in Man.

Current Practice

The clinical setting in which patients are evaluated determines the scope of practice, diagnostic, and therapeutic interventions. For the purposes of general discussion, the typical encounters between patients and genetic practitioners may involve:

- Referral to an out-patient genetics clinic (pediatric, adult, or combined) or an in-hospital consultation, most often for diagnostic evaluation.
- Specialty genetics clinics focusing on management of inborn errors of metabolism, skeletal dysplasia, or lysosomal storage diseases.
- Referral for counseling in a prenatal genetics clinic to discuss risks to the pregnancy (advanced maternal age, teratogen exposure, family history of a genetic disease), test results (abnormal maternal serum screen, abnormal ultrasound), and/or options for prenatal diagnosis (typically amniocentesis or chorionic villus sampling).

- Multidisciplinary specialty clinics that include a clinical geneticist or genetic counselor (cancer genetics, cardiovascular genetics, craniofacial or cleft lip/palate, hearing loss clinics, muscular dystrophy/ neurodegenerative disorder clinics).

Diagnostic Evaluation

Each patient will undergo a diagnostic evaluation tailored to their own particular presenting signs and symptoms. The geneticist will establish a differential diagnosis and recommend appropriate testing. Increasingly, clinicians use SimulConsult, paired with the National Library of Medicine Gene Review articles, to narrow the list of hypotheses (known as the differential diagnosis) and identify the tests that are relevant for a particular patient. These tests might evaluate for chromosomal disorders, inborn errors of metabolism, or single gene disorders.

Chromosome Studies

Chromosome studies are used in the general genetics clinic to determine a cause for developmental delay/mental retardation, birth defects, dysmorphic features, and/or autism. Chromosome analysis is also performed in the prenatal setting to determine whether a fetus is affected with aneuploidy or other chromosome rearrangements. Finally, chromosome abnormalities are often detected in cancer samples. A large number of different methods have been developed for chromosome analysis:

- Chromosome analysis using a karyotype involves special stains that generate light and dark bands, allowing identification of each chromosome under a microscope.
- Fluorescence in situ hybridization (FISH) involves fluorescent labeling of probes that bind to specific DNA sequences, used for identifying aneuploidy, genomic deletions or duplications, characterizing chromosomal translocations and determining the origin of ring chromosomes.
- Chromosome painting is a technique that uses fluorescent probes specific for each chromosome to differentially label each chromosome. This technique is more often used in cancer cytogenetics, where complex chromosome rearrangements can occur.
- Array comparative genomic hybridization is a new molecular technique that involves hybridization of an individual DNA sample to a glass slide or microarray chip containing molecular probes (ranging from large ~200kb bacterial artificial chromosomes to small oligonucleotides) that represent unique regions of the genome. This method is particularly sensitive for detection of genomic gains or losses across the genome but

does not detect balanced translocations or distinguish the location of duplicated genetic material (for example, a tandem duplication versus an insertional duplication).

Basic Metabolic Studies

Biochemical studies are performed to screen for imbalances of metabolites in the bodily fluid, usually the blood (plasma/serum) or urine, but also in cerebrospinal fluid (CSF). Specific tests of enzyme function (either in leukocytes, skin fibroblasts, liver, or muscle) are also employed under certain circumstances. In the US, the newborn screen incorporates biochemical tests to screen for treatable conditions such as galactosemia and phenylketonuria (PKU).

Patients suspected to have a metabolic condition might undergo the following tests:

- Quantitative amino acid analysis is typically performed using the ninhydrin reaction, followed by liquid chromatography to measure the amount of amino acid in the sample (either urine, plasma/serum, or CSF). Measurement of amino acids in plasma or serum is used in the evaluation of disorders of amino acid metabolism such as urea cycle disorders, maple syrup urine disease, and PKU. Measurement of amino acids in urine can be useful in the diagnosis of cystinuria or renal Fanconi syndrome as can be seen in cystinosis.
- Urine organic acid analysis can be either performed using quantitative or qualitative methods, but in either case the test is used to detect the excretion of abnormal organic acids. These compounds are normally produced during bodily metabolism of amino acids and odd-chain fatty acids, but accumulate in patients with certain metabolic conditions.
- The acylcarnitine combination profile detects compounds such as organic acids and fatty acids conjugated to carnitine. The test is used for detection of disorders involving fatty acid metabolism, including MCAD.
- Pyruvate and lactate are byproducts of normal metabolism, particularly during anaerobic metabolism. These compounds normally accumulate during exercise or ischemia, but are also elevated in patients with disorders of pyruvate metabolism or mitochondrial disorders.
- Ammonia is an end product of amino acid metabolism and is converted in the liver to urea through a series of enzymatic reactions termed the urea cycle. Elevated ammonia can therefore be detected in patients with urea cycle disorders, as well as other conditions involving liver failure.
- Enzyme testing is performed for a wide range of metabolic disorders to confirm a diagnosis suspected based on screening tests.

Molecular Studies

- DNA sequencing is used to directly analyze the genomic DNA sequence of a particular gene. In general, only the parts of the gene that code for the expressed protein (exons) and small amounts of the flanking untranslated regions and introns are analyzed. Therefore, although these tests are highly specific and sensitive, they do not routinely identify all of the mutations that could cause disease.
- DNA methylation analysis is used to diagnose certain genetic disorders that are caused by disruptions of epigenetic mechanisms such as genomic imprinting and uniparental disomy.
- Southern blotting is an early technique basic on detection of fragments of DNA separated by size through gel electrophoresis and detected using radiolabeled probes. This test was routinely used to detect deletions or duplications in conditions such as Duchenne muscular dystrophy but is being replaced by high-resolution array comparative genomic hybridization techniques. Southern blotting is still useful in the diagnosis of disorders caused by trinucleotide repeats.
- Short tandem repeats are unique markers that can be used to determine haplotypes and are used in identity testing for maternal cell contamination.

Treatments

Each cell of the body contains the hereditary information (DNA) wrapped up in structures called chromosomes. Since genetic syndromes are typically the result of alterations of the chromosomes or genes, there is no treatment currently available that can correct the genetic alterations in every cell of the body. Therefore, there is currently no "cure" for genetic disorders. However, for many genetic syndromes there is treatment available to manage the symptoms. In some cases, particularly inborn errors of metabolism, the mechanism of disease is well understood and offers the potential for dietary and medical management to prevent or reduce the long-term complications. In other cases, infusion therapy is used to replace the missing enzyme. Current research is actively seeking to use gene therapy or other new medications to treat specific genetic disorders.

Management of Metabolic Disorders

In general, metabolic disorders arise from enzyme deficiencies that disrupt normal metabolic pathways. Compound "A" is metabolized to "B" by enzyme "X", compound "B" is metabolized to "C" by enzyme "Y", and compound "C" is metabolized to "D" by enzyme "Z". If enzyme "Z" is missing, compound "D" will be missing, while compounds "A", "B", and "C" will build up. The pathogenesis of this particular condition could result from lack of compound "D", if it is critical for some cellular function, or from toxicity due to excess "A", "B", and/or "C". Treatment of the metabolic disorder could be achieved through dietary supplementation of compound "D" and dietary restriction of compounds "A", "B", and/or "C" or by treatment with a medication that promoted disposal of excess "A", "B", or "C". Another approach that can be taken is enzyme replacement therapy, in which a patient is given an infusion of the missing enzyme.

Diet

Dietary restriction and supplementation are key measures taken in several well-known metabolic disorders, including galactosemia, phenylketonuria (PKU), maple syrup urine disease, organic acidurias and urea cycle disorders. Such restrictive diets can be difficult for the patient and family to maintain, and require close consultation with a nutritionist who has special experience in metabolic disorders. The composition of the diet will change depending on the caloric needs of the growing child and special attention is needed during a pregnancy if a woman is affected with one of these disorders.

Medication

Medical approaches include enhancement of residual enzyme activity (in cases where the enzyme is made but is not functioning properly), inhibition of other enzymes in the biochemical pathway to prevent buildup of a toxic compound, or diversion of a toxic compound to another form that can be excreted. Examples include the use of high doses of pyridoxine (vitamin B6) in some patients with homocystinuria to boost the activity of the residual cystathione synthase enzyme, administration of biotin to restore activity of several enzymes affected by deficiency of biotinidase, treatment with NTBC in Tyrosinemia to inhibit the production of succinylacetone which causes liver toxicity, and the use of sodium benzoate to decrease ammonia build-up in urea cycle disorders.

Enzyme Replacement Therapy

Certain lysosomal storage diseases are treated with infusions of a recombinant enzyme (produced in a laboratory), which can reduce the accumulation of the compounds in various tissues. Examples include Gaucher disease, Fabry disease, Mucopolysaccharidoses and Glycogen storage disease type II. Such treatments are limited by the ability of the enzyme to reach the affected areas (the blood brain barrier prevents enzyme from reaching the brain, for example), and can sometimes be associated with allergic reactions. The long-term clinical effectiveness of enzyme replacement therapies vary widely among different disorders.

Other Examples

- Angiotensin receptor blockers in Marfan syndrome & Loeys-Dietz
- Bone marrow transplantation
- Gene therapy

Ethical, Legal and Social Implications

Genetic information provides a unique type of knowledge about an individual and his/her family, fundamentally different than a typically laboratory test that provides a "snapshot" of an individual's health status. The unique status of genetic information and inherited disease has a number of ramifications with regard to ethical, legal, and societal concerns.

Societies

The more empirical approach to human and medical genetics was formalized by the founding in 1948 of the American Society of Human Genetics. The Society first began annual meetings that year (1948) and its international counterpart, the International Congress of Human Genetics, has met every 5 years since its inception in 1956. The Society publishes the American Journal of Human Genetics on a monthly basis.

Medical genetics is now recognized as a distinct medical specialty in the U.S. with its own approved board (the American Board of Medical Genetics) and clinical specialty college (the American College of Medical Genetics). The College holds an annual scientific meeting, publishes a monthly journal, Genetics in Medicine, and issues position papers and clinical practice guidelines on a variety of topics relevant to human genetics.

Research

This section requires expansion with:

The broad range of research in medical genetics reflects the overall scope of this field, including basic research on genetic inheritance and the human genome, mechanisms of genetic and metabolic disorders, translational research on new treatment modalities, and the impact of genetic testing

Basic Genetics Research

Basic research geneticists usually undertake research in universities, biotechnology firms and research institutes.

Allelic Architecture of Disease

Sometimes the link between a disease and an unusual gene variant is more subtle. The genetic architecture of common diseases is an important factor in determining the extent to which patterns of genetic variation influence group differences in health outcomes. According to the common disease/common variant hypothesis, common variants present in the ancestral population before the dispersal of modern humans from Africa play an important role in human diseases. Genetic variants associated with Alzheimer disease, deep venous thrombosis, Crohn disease, and type 2 diabetes appear to adhere to this model. However, the generality of the model has not yet been established and, in some cases, is in doubt. Some diseases, such as many common cancers, appear not to be well described by the common disease/common variant model.

Another possibility is that common diseases arise in part through the action of combinations of variants that are individually rare. Most of the disease-associated alleles discovered to date have been rare, and rare variants are more likely than common variants to be differentially distributed among groups distinguished by ancestry. However, groups could harbor different, though perhaps overlapping, sets of rare variants, which would reduce contrasts between groups in the incidence of the disease.

The number of variants contributing to a disease and the interactions among those variants also could influence the distribution of diseases among groups. The difficulty that has been encountered in finding contributory alleles for complex diseases and in replicating positive associations suggests that many complex diseases involve numerous variants rather than a moderate number of alleles, and the influence of any given variant may depend in critical ways on the genetic and environmental background. If many alleles are required to increase susceptibility to a disease, the odds are low that the necessary combination of alleles would become concentrated in a particular group purely through drift.

Population Substructure in Genetics Research

One area in which population categories can be important considerations in genetics research is in controlling for confounding between population substructure, environmental exposures, and health outcomes.

Association studies can produce spurious results if cases and controls have differing allele frequencies for genes that are not related to the disease being studied, although the magnitude of this problem in genetic association studies is subject to debate. Various methods have been developed to detect and account for population substructure, but these methods can be difficult to apply in practice.

Population substructure also can be used to advantage in genetic association studies. For example, populations that represent recent mixtures of geographically separated ancestral groups can exhibit longer-range linkage disequilibrium between susceptibility alleles and genetic markers than is the case for other populations. Genetic studies can use this admixture linkage disequilibrium to search for disease alleles with fewer markers than would be needed otherwise. Association studies also can take advantage of the contrasting experiences of racial or ethnic groups, including migrant groups, to search for interactions between particular alleles and environmental factors that might influence health.

Human Genetic Clustering

Human genetic clustering analysis uses mathematical cluster analysis of the degree of similarity of genetic data between individuals and groups to infer population structures and assign individuals to groups that often correspond with their self-identified geographical ancestry. A similar analysis can be done using principal components analysis, which in earlier research was a popular method. Many of recent studies in the past few years have returned to using principal components analysis.

Studies

Phylogenetic tree by Cavalli-Sforza & al. (1994)

Linkage tree for 9 population clusters showing genetic distances (F_{ST})
(Cavalli-Sforza *et al.*, 1994:80)

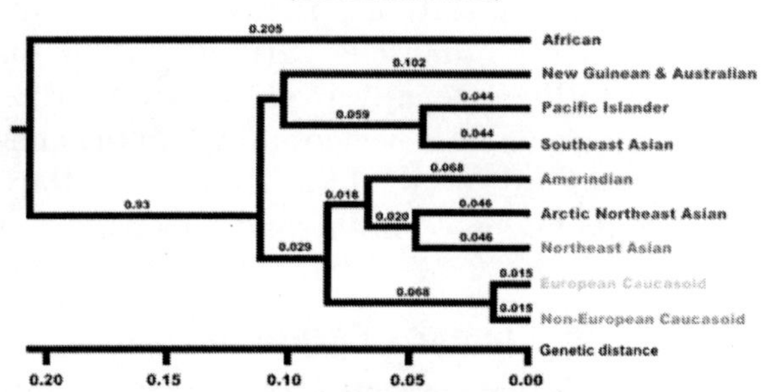

F_{ST} distance matrix for the 9 clusters shown above
(x10,000 with standard errors obtained by bootstrap analysis)

	AFR	NEC	EUC	NEA	ANE	AME	SEA	PAI	NGA
African	0.0								
Non-European Caucasian	1340.0 ± 301	0.0							
European Caucasian	1656.6 ± 416	154.7 ± 29	0.0						
Northeast Asian	1979.1 ± 452	640.4 ± 134	938.2 ± 217	0.0					
Arctic Northeast Asian	2008.5 ± 387	708.2 ± 160	746.7 ± 210	459.7 ± 98	0.0				
Amerindian	2261.4 ± 434	955.5 ± 204	1038.2 ± 276	746.5 ± 183	577.4 ± 89	0.0			
Southeast Asian	2206.3 ± 529	939.6 ± 262	1240.4 ± 339	630.5 ± 299	1039.4 ± 326	1341.7 ± 418	0.0		
Pacific Islander	2505.4 ± 648	953.7 ± 230	1344.7 ± 354	723.8 ± 262	1181.2 ± 331	1740.7 ± 544	436.7 ± 87	0.0	
New Guinean and Australian	2472.0 ± 536	1179.1 ± 189	1345.7 ± 231	734.4 ± 118	1012.5 ± 257	1457.9 ± 263	1237.9 ± 277	808.7 ± 264	0.0

Neighbor-joining method, by Naruya Saitou and Masatoshi Nei (2002)

In 2004, Lynn Jorde and Steven Wooding argued that "Analysis of many loci now yields reasonably accurate estimates of genetic similarity among individuals, rather than populations. Clustering of individuals is correlated with geographic origin or ancestry."

Gene clusters from Rosenberg (2006) for K=7 clusters. (Cluster analysis divides a dataset into any prespecified number of clusters.) Individuals have genes from multiple clusters. The cluster prevalent only among the Kalash people (yellow) only splits off at K=7 and greater.

Human population structure can be inferred from multilocus DNA sequence data (Rosenberg et al. 2002, 2005). Individuals from 52 populations were examined at 993 DNA markers. This data was used to partition individuals into K = 2, 3, 4, 5, or 6 gene clusters. In this figure, the average fractional membership of individuals from each population is represented by horizontal bars partitioned into K colored segments.

A study by Neil Risch in 2005 used 326 microsatellite markers and self-identified race/ethnic group (SIRE), white (European American), African-American (black), Asian and Hispanic (individuals involved in the study had to choose from one of these categories), to representing discrete "populations", and showed distinct and non-overlapping clustering of the white,

African-American and Asian samples. The results were claimed to confirm the integrity of self-described ancestry: "We have shown a nearly perfect correspondence between genetic cluster and SIRE for major ethnic groups living in the United States, with a discrepancy rate of only 0.14%."(Tang, 2005)

Studies such as those by Risch and Rosenberg use a computer program called STRUCTURE to find human populations (gene clusters). It is a statistical program that works by placing individuals into one of two clusters based on their overall genetic similarity, many possible pairs of clusters are tested per individual to generate multiple clusters. These populations are based on multiple genetic markers that are often shared between different human populations even over large geographic ranges. The notion of a genetic cluster is that people within the cluster share on average similar allele frequencies to each other than to those in other clusters. (A. W. F. Edwards, 2003 but also infobox "Multi Locus Allele Clusters") In a test of idealised populations, the computer programme STRUCTURE was found to consistently under-estimate the numbers of populations in the data set when high migration rates between populations and slow mutation rates (such as single-nucleotide polymorphisms) were considered.

Nevertheless the Rosenberg et al. (2002) paper shows that individuals can be assigned to specific clusters to a high degree of accuracy. One of the underlying questions regarding the distribution of human genetic diversity is related to the degree to which genes are shared between the observed clusters. It has been observed repeatedly that the majority of variation observed in the global human population is found within populations. This variation is usually calculated using Sewall Wright's Fixation index (F_{ST}), which is an estimate of between to within group variation. The degree of human genetic variation is a little different depending upon the gene type studied, but in general it is common to claim that ~85% of genetic variation is found within groups, ~6–10% between groups within the same continent and ~6–10% is found between continental groups.

For example The Human Genome Project states "two random individuals from any one group are almost as different [genetically] as any two random individuals from the entire world." Sarich and Miele, however, have argued that estimates of genetic difference between individuals of different populations fail to take into account human diploidity.

The point is that we are diploid organisms, getting one set of chromosomes from one parent and a second from the other. To the extent that your mother and father are not especially closely related, then, those two sets of chromosomes will come close to being a random sample of the chromosomes in your population. And the sets present in some randomly chosen member of yours will also be about as different from your two sets as they are from one another. So how much of the variability will be distributed where? First is the 15 percent that is interpopulational.

The other 85 percent will then split half and half (42.5 percent) between the intra- and interindividual within-population comparisons. The increase in variability in between-population comparisons is thus 15 percent against the 42.5 percent that is between-individual within-population. Thus, 15/42.5 is 32.5 percent, a much more impressive and, more important, more legitimate value than 15 percent.

Additionally, Edwards (2003) claims in his essay "Lewontin's Fallacy" that: "It is not true, as Nature claimed, that 'two random individuals from any one group are almost as different as any two random individuals from the entire world'" and Risch et al. (2002) state "Two Caucasians are more similar to each other genetically than a Caucasian and an Asian." It should be noted that these statements are not the same. Risch et al.

Simply state that two indigenous individuals from the same geographical region are more similar to each other than either is to an indigenous individual from a different geographical region, a claim few would argue with. Jorde et al. put it like this:

The picture that begins to emerge from this and other analyses of human genetic variation is that variation tends to be geographically structured, such that most individuals from the same geographic region will be more similar to one another than to individuals from a distant region.

Whereas Edwards claims that it is not true that the differences between individuals from different geographical regions represent only a small proportion of the variation within the human population (he claims that within group differences between individuals are not almost as large as between group differences). Bamshad et al. (2004) used the data from Rosenberg et al.

(2002) to investigate the extent of genetic differences between individuals within continental groups relative to genetic differences between individuals between continental groups. They found that though these individuals could be classified very accurately to continental clusters, there was a significant degree of genetic overlap on the individual level, to the extent that, using 377 loci, individual Europeans were about 38% of the time more genetically similar to East Asians than to other Europeans.

The existence of allelic clines and the observation that the bulk of human variation is continuously distributed, has led some scientists to conclude that any categorization schema attempting to partition that variation meaningfully will necessarily create artificial truncations. (Kittles & Weiss 2003).

It is for this reason, Reanne Frank argues, that attempts to allocate individuals into ancestry groupings based on genetic information have yielded varying results that are highly dependent on methodological design. Serre and Pääbo (2004) make a similar claim:

The absence of strong continental clustering in the human gene pool is of practical importance. It has recently been claimed that "the greatest genetic structure that exists in the human population occurs at the racial level" (Risch et al. 2002). Our results show that this is not the case, and we see no reason to assume that "races" represent any units of relevance for understanding human genetic history.

In a response to Serre and Pääbo (2004), Rosenberg et al. (2005) make three relevant observations. Firstly they maintain that their clustering analysis is robust. Secondly they agree with Serre and Pääbo that membership of multiple clusters can be interpreted as evidence for clinality (isolation by distance), though they also comment that this may also be due to admixture between neighbouring groups (small island model). Thirdly they comment that evidence of clusterdness is not evidence for any concepts of "biological race".

Risch et al. (2002) state that "two Caucasians are more similar to each other genetically than a Caucasian and an Asian", but Bamshad et al. (2004) used the same data set as Rosenberg et al. (2002) to show that Europeans are more similar to Asians 38% of the time than they are to other Europeans when only 377 microsatellite markers are analysed. If a landmass is considered with variation distributed in one dimension (west-east). Top: Distribution of genetic variation if a small island model is considered; there are two "populations" with a narrow region of hybridisation where migration occurs. This pattern is "clustered".

Bottom: Distribution of genetic variation if isolation by distance is considered; all variation is gradual over the extent of the landmass. This pattern is "clinal".

Percentage similarity between two individuals from different clusters when 377 microsatellite markers are considered.

x	***Africans***	***Europeans***	***Asians***
Europeans	36.5	—	—
Asians	35.5	38.3	—
Indigenous Americans	26.1	33.4	35

In agreement with the observation of Bamshad et al. (2004), Witherspoon et al. (2007) have shown that many more than 326 or 377 microsatellite loci are required in order to show that individuals are always more similar to individuals in their own population group than to individuals in different population groups, even for three distinct populations.

Witherspoon et al. (2007) have argued that even when individuals can be reliably assigned to specific population groups, it may still be possible for two randomly chosen individuals from different populations/clusters to be more similar to each other than to a randomly chosen member of their own cluster.

They found that many thousands of genetic markers had to be used in order for the answer to the question "How often is a pair of individuals from one population genetically more dissimilar than two individuals chosen from two different populations?" to be "never". This assumed three population groups separated by large geographic ranges (European, African and East Asian).

The entire world population is much more complex and studying an increasing number of groups would require an increasing number of markers for the same answer.

Witherspoon et al. conclude that "caution should be used when using geographic or genetic ancestry to make inferences about individual phenotypes."

Clustering does not particularly correspond to continental divisions. Depending on the parameters given to their analytical program, Rosenberg and Pritchard were able to construct between divisions of between 4 and 20 clusters of the genomes studied, although they excluded analysis with more than 6 clusters from their published article. Probability values for various cluster configurations varied widely, with the single most likely configuration coming with 16 clusters although other 16-cluster configurations had low probabilities. Overall, "there is no clear evidence that K=6 was the best estimate" according to geneticist Deborah Bolnick (2008:76-77).

Population Groups in Biomedicine

Race and Health

Race and health research, often done in the United States, has found both current and historical racial differences in the frequency, treatments, and availability of treatments for several diseases. This can add up to significant group differences in variables such as life expectancy.

Many explanations for such differences have been argued, including socioeconomic factors (e.g., education, employment, and income), lifestyle behaviors (e.g., physical activity and alcohol intake), social environment (e.g., educational and economic opportunities, racial/ethnic discrimination, and neighborhood and work conditions), and access to preventive health-care services (e.g., cancer screening and vaccination) as well as to treatment (through lack of insurance, lack of hospitals in certain areas, etc.), among other environmental differences. Some diseases may also be influenced by genes which differ in frequency between groups, such as sickle-cell anemia, which occurs overwhelmingly among some black populations, although the significance in clinical medicine of race categories as a proxy for exact genotypes of individuals has been questioned.

Health Disparities

Health disparities refer to gaps in the quality of health and health care across racial and ethnic groups. The US Health Resources and Services Administration defines health disparities as "population-specific differences in the presence of disease, health outcomes, or access to health care." Health is measured through variables such as life expectancy and incidence of diseases.

Single Gene Disorders

There are many single gene genetic disorders that differ in frequency between different popultions. Some examples are discussed below.

Cystic fibrosis is the most common life-limiting autosomal recessive disease among people of European heritage.

Sickle-cell anemia is most prevalent in populations of sub-Saharan African ancestry, but it is also common among Latin-American, Indian, Saudi Arab, and Mediterranean populations such as Turkey, Greece, and Italy. Tay-Sachs Disease is an autosomal recessive disorder that is more frequent among Ashkenazi Jews than among other Jewish groups and non-Jewish populations.

Hereditary hemochromatosis is most common among those of Northern European ancestry, in particular those of Celtic descendent. Lactose intolerance is another examples of single gene genetic disorders that differ in frequency between populations.

Multifactorial Polygenic Diseases

These are many diseases which differ in frequency between different populations.

Disease Progression

Groups may differ in how a disease progresses.

Treatment

Groups may differ in their responses to treatment.

Medicines Targeted at Specific Races

Race-based medicine is the term for medicines that are targeted at specific ethnic clusters which are shown to have a propensity for a certain disorder.

The first example of this in the U.S. was when BiDil, a medication for congestive heart failure, was licensed specifically for use in American patients that self-identify as black. Previous studies had shown that African American patients with congestive heart failure generally respond less effectively to traditional treatments than white patients with similar conditions.

After two trials, BiDil was licensed exclusively for use in African American patients. Critics have argued that this particular licensing was unwarranted, since the trials did not in fact show that the drug was more effective in African Americans than in other groups, but merely that it was more effective in African Americans thanother similar drug.

It was also only tested in African American males, but not in any other racial groups or among women. This peculiar trial and licensing procedure has prompted suggestions that the licensing was in fact used as a race based advertising scheme.

Environmental Explanations

In multiracial societies such as the United States, racial groups differ greatly in regard to social and cultural factors such as socioeconomic status, healthcare, diet, and education.

There is also the presence of racism which some see as a very important explaining factor. Some argue that for many diseases racial differences would disappear if all environmental factors could be controlled for. Article about race and health in the United States for a discussion of such factors. These factors may or may not be appropriate in other nations.

Ethnic minorities may also have specific health care needs which need to be taken into consideration by health services in order to tackle health disparities.

Genetic Explanations

Evolutionary Explanations

Genes may be under strong selection in response to local diseases. For example, people who are duffy negative tend to have higher resistance to malaria. Most Africans are duffy negative and most non-Africans are duffy positive. A number of genetic diseases more prevalent in malaria-afflicted areas may provide some genetic resistance to malaria including sickle cell disease, thalassaemias, glucose-6-phosphate dehydrogenase, and possibly others.

Many theories about the origin of the Cystic fibrosis have suggested that it provides a heterozygote advantage by giving resistance to diseases earlier common in Europe.

In earlier research a common theory was the "common disease-common variant" model. It argues that for common illnesses, the genetic contribution comes from the additive or multiplicative effects of gene variants that each one is common in the population. Each such gene variant is argued to cause only a small risk of disease and no single variant is enough to cause the disease. An individual must have many of these common gene variants in order for the risk of disease to be substantial.

More recent research indicates that the "common disease-rare variant" may be a better explanation for many common diseases. In this model, rare but higher-risk gene variants cause common diseases. This model may be particularly relevant for diseases that reduces fertility.

In contrast, for common genes associated with common disease to persist they must either have little effect during the reproductive period of life (like Alzheimer's disease) or provide some advantage in the original environment (like genes causing autoimmune diseases also providing resistance against infections). In either case varying frequencies of genes variants in different populations may be an explanation for health disparities. Genetic variants associated with Alzheimer's disease, deep venous thrombosis, Crohn disease, and type 2 diabetes appear to adhere to "common disease-common variant" model.

Gene Flow

Gene flow and admixture can also have an effect on relationships between race and race-linked disorders. Multiple sclerosis, for example, is typically associated with people of European descent, but due to admixture African Americans have elevated levels of the disorder relative to Africans. Some diseases and physiological variables vary depending upon their admixture ratios. Examples include measures of insulin functioning and obesity.

Gene Interactions

The same gene variant, or group of gene variants, may produce different effects in different populations depending on differences in the gene variants, or groups of gene variants, they interact with. One example is the rate of progression to AIDS and death in HIV–infected patients.

In Caucasians and Hispanics, HHC haplotypes were associated with disease retardation, particularly a delayed progression to death. In contrast, for African Americans, possession of HHC haplotypes was associated with disease acceleration.

Controversy Regarding Race in Biomedicine

There is a controversy regarding race as a method for classifying humans. Some it as a purely social construct. Others it as a biological reality reflecting average genetic group differences.

New interest in human biological variation has resulted in a resurgence of the use of race in biomedicine.

The main impetus for this development is the possibility of improving the prevention and treatment of certain diseases by predicting hard-to-ascertain factors, such as genetically conditioned health factors, on the basis of more easily ascertained characteristics such as phenotype and racial self-identification.

Since medical judgment often involves decision-making under uncertain conditions, many doctors consider it useful to take race into account when treating disease because diseases and treatment responses tend to cluster by geographic ancestry. The discovery that more diseases than previously thought correlate with racial identification have further sparked the interest in using race as a proxy for bio-geographical ancestry and genetic buildup.

Race in medicine is used as an approximation for more specific genetic and environmental risk factors. Race is thus partly a surrogate for environmental factors such as differences in socioeconomic status that are known to affect health.

It is also an imperfect surrogate for ancestral geographic regions and differences in gene frequencies between different ancestral populations and thus differences in genes that can affect health. This can give an approximation of probability for disease or for preferred treatment, although the approximation is less than perfect.

Taking the example of sickle-cell disease, in an emergency room, knowing the geographic origin of a patient may help a doctor doing an initial diagnosis if a patient presents with symptoms compatible with this disease. This is unreliable evidence with the disease being present in many different groups as noted above with the trait also present in some Mediterranean European populations. Definitive diagnosis comes from examining the blood of the patient. In the US, screening for sickle cell anemia is done on all newborns regardless of race.

The continued use of racial categories has been criticized. Apart from the general controversy regarding race, some argue that the continued use of racial categories in health care and as risk factors could result in increased stereotyping and discrimination in society and health services. On the other hand, also some of those who are critical of race as a biological concept see race as socially meaningful group that is important to study epidemiologically in order to reduce disparities. David Williams (1994) argued, after an examination of articles in the journal Health Services Research during the 1966-90 period, that how race was determined and defined was seldom described. At a minimum, researchers should describe if race was assessed by self-report, proxy report, extraction from records, or direct observation. Race was also often used questionable, such as an indicator of socioeconomic status. Racial genetic explanations may be overemphasized, ignoring the interaction with and the role of the environment.

There is general agreement that a goal of health-related genetics should be to move past the weak surrogate relationships of racial health disparity and get to the root causes of health and disease. This includes research which strives to analyze human genetic variation in smaller groups than races across the world. One such method is called ethnogenetic layering. It works by focusing on geographically identified microethnic groups. For example, in the Mississippi Delta region ethnogenetic layering might include such microethnic groups as the Cajun (as a subset of European Americans), the Creole and Black groups [with African origins in Senegambia, Central Africa and Bight of Benin] (as a subset of African Americans), and Choctaw, Houmas, Chickasaw, Coushatta, Caddo, Atakapa, Karankawa and Chitimacha peoples (as subsets of Native American Indians).

Better still may be individual genetic assessment of relevant genes. However, until cheaper and more widely available methods of genetic analysis are commonplace, the consideration of race remains a worthwhile practice for many doctors and researchers. Even when such methods become commonly available, race will continue to be important when looking at groups instead of individuals such as in epidemiologic research.

Association Studies

One area in which population categories can be important considerations in genetics research is in controlling for confounding between population genetic substructure, environmental exposures, and health outcomes. Association studies can produce spurious results if cases and controls have differing allele frequencies for genes that are not related to the disease being studied, although the magnitude of its problem in genetic association studies is subject to debate. Various techniques detect and account for population substructure, but these methods can be difficult to apply in practice.

Population genetic substructure also can aid genetic association studies. For example, populations that represent recent mixtures of separated ancestral groups can exhibit longer-range linkage disequilibrium between susceptibility alleles and genetic markers than is the case for other populations. Genetic studies can use this disequilibrium to search for disease alleles with fewer markers than would be needed otherwise. Association studies also can take advantage of the contrasting experiences of racial or ethnic groups, including migrant groups, to search for interactions between particular alleles and environmental factors that might influence health.

Human Genome Projects

The Human Genome Diversity Project has collected genetic samples from 52 indigenous populations. This sampling has been controversial because some fear that racists might use the results, because the sampling might overemphasize genetic distinctions, because of informed consent issues, and patenting issues.

Race and Health in the United States

Research on Race and health in the United States shows many health disparities between the different racial/ethnic groups. The possible causes, such as genetics, socioeconomic factors, and racism, continue to be debated.

Southern Blot

A Southern blot is a method routinely used in molecular biology for detection of a specific DNA sequence in DNA samples. Southern blotting combines transfer of electrophoresis-separated DNA fragments to a filter membrane and subsequent fragment detection by probe hybridization. The method is named after its inventor, the British biologist Edwin Southern. Other blotting methods (i.e., Western blot, Northern blot, Eastern blot, Southwestern blot) that employ similar principles, but using RNA or protein, have later been named in reference to Edwin Southern's name. As the technique was eponymously named, Southern blot is capitalized as is conventional for proper nouns. The names for other blotting methods may follow this convention, by analogy. Method

1. Restriction endonucleases are used to cut high-molecular-weight DNA strands into smaller fragments.
2. The DNA fragments are then electrophoresed on an agarose gel to separate them by size.

3. If some of the DNA fragments are larger than 15 kb, then prior to blotting, the gel may be treated with an acid, such as dilute HCl, which depurinates the DNA fragments, breaking the DNA into smaller pieces, thus allowing more efficient transfer from the gel to membrane.
4. If alkaline transfer methods are used, the DNA gel is placed into an alkaline solution (typically containing sodium hydroxide) to denature the double-stranded DNA. The denaturation in an alkaline environment may improve binding of the negatively charged DNA to a positively charged membrane, separating it into single DNA strands for later hybridization to the probe, and destroys any residual RNA that may still be present in the DNA. The choice of alkaline over neutral transfer methods, however, is often empirical and may result in equivalent results.
5. A sheet of nitrocellulose (or, alternatively, nylon) membrane is placed on top of (or below, depending on the direction of the transfer) the gel. Pressure is applied evenly to the gel (either using suction, or by placing a stack of paper towels and a weight on top of the membrane and gel), to ensure good and even contact between gel and membrane. If transferring by suction 20X SSC buffer is used to ensure a seal and prevent drying of the gel. Buffer transfer by capillary action from a region of high water potential to a region of low water potential (usually filter paper and paper tissues) is then used to move the DNA from the gel on to the membrane; ion exchange interactions bind the DNA to the membrane due to the negative charge of the DNA and positive charge of the membrane.
6. The membrane is then baked in a vacuum or regular oven at 80 °C for 2 hours (standard conditions; nitrocellulose or nylon membrane) or exposed to ultraviolet radiation (nylon membrane) to permanently attach the transferred DNA to the membrane.
7. The membrane is then exposed to a hybridization probe—a single DNA fragment with a specific sequence whose presence in the target DNA is to be determined. The probe DNA is labelled so that it can be detected, usually by incorporating radioactivity or tagging the molecule with a fluorescent or chromogenic dye. In some cases, the hybridization probe may be made from RNA, rather than DNA. To ensure the specificity of the binding of the probe to the sample DNA, most common hybridization methods use salmon or herring sperm DNA for blocking of the membrane surface and target DNA, deionized formamide, and detergents such as SDS to reduce non-specific binding of the probe.
8. After hybridization, excess probe is washed from the membrane (typically using SSC buffer), and the pattern of hybridization is visualized on X-ray film by autoradiography in the case of a radioactive or fluorescent probe, or by development of color on the membrane if a chromogenic detection method is used.

Result

Hybridization of the probe to a specific DNA fragment on the filter membrane indicates that this fragment contains DNA sequence that is complementary to the probe. The transfer step of the DNA from the electrophoresis gel to a membrane permits easy binding of the labeled hybridization probe to the size-fractionated DNA. It also allows for the fixation of the target-probe hybrids, required for analysis by autoradiography or other detection methods. Southern blots performed with restriction enzyme-digested genomic DNA may be used to determine the number of sequences (e.g., gene copies) in a genome.

A probe that hybridizes only to a single DNA segment that has not been cut by the restriction enzyme will produce a single band on a Southern blot, whereas multiple bands will likely be observed when the probe hybridizes to several highly similar sequences (e.g., those that may be the result of sequence duplication).

Modification of the hybridization conditions (for example, increasing the hybridization temperature or decreasing salt concentration) may be used to increase specificity and decrease hybridization of the probe to sequences that are less than 100% similar.

Triple-stranded DNA

A triple-stranded DNA is a structure of DNA in which three oligonucleotides wind around each other and form a triple helix. In this structure, one strand binds to a B-form DNA double helix through Hoogsteen or reversed Hoogsteen hydrogen bonds.

For example, a nucleobase T binds to a Watson–Crick base-pairing of T-A by Hoogsteen hydrogen bonds between an AxT pair (x represents a Hoogsteen base pair). An N-3 protonated cytosine, represented as C+, can also form a base-triplet with a C-G pair through the Hoogsteen base-pairing of an GxC+. Thus, the triple-helical DNAs using these Hoogsteen pairings consist of two homopyrimidines and one homopurine, and the homopyrimidine third strand is parallel to the homopurine strand.

A homopurine third strand can also bind to a homopurine-homopyrimidine duplex using reversed Hoogsteen patterns. In this triplex, a nucleobase A binds to a T-A base pair and a G to a C-G pair. Since the nucleobases on the third strand have to be reversed, the homopurine third strand is antiparallel to the homopurine strand of the original duplex.

Triple-stranded DNA was a common hypothesis in the 1950s when scientists were struggling to discover DNA's true structural from. Watson and Crick (who later won the Nobel Prize for their double-helix model) originally considered a triple-helix model, as did Pauling and Corey, who published a proposal for their triple-helix model in the 1953 scientific journal Nature, as well as fellow scientist Fraser. However, Watson and Crick soon identified several problems with these models:

1. Negatively charged phosphates near the axis will repel each other, leaving the question as to how the three-chain structure would stay together.
2. In a triple-helix model (specifically Pauling and Corey's model), some of the van der Waals distances appear to be too small.

Fraser's model differed from Pauling and Corey's in that in his model the phosphates are on the outside and the bases are on the inside, linked together by hydrogen bonds.

However, Watson and Crick found Fraser's model to be too ill-defined to comment specifically on its inadequacies in their publication in "Nature" (1953): Molecular Structure of Nucleic Acids. Triple-stranded DNA was also described in 1957, when it was thought to occur in only one in vivo biological process: as an intermediate product during the action of the E. coli recombination enzyme RecA. Its role in that process is not understood.

Using nucleic acid segments that bind to the DNA duplexes to form triple strands as a way of regulating gene expression is under investigation by biotechnology companies. Similar work is also being undertaken at Yale University.

Triple-stranded DNA has been implicated in the regulation of several genes. One, the c-myc gene has been extensively mutated to examine the exact role that triplex DNA structure, versus linear sequence, plays in gene regulation. A c-myc promoter element, termed the nuclease-sensitive element or NSE, can form tandem intramolecular triplexes of the H-DNA type and has a repeating sequence motif (ACCCTCCCC)4.

The NSE when mutated and examined for transcriptional activity and for intra- and intermolecular triplex forming ability. The transcriptional activity of mutant NSEs can be predicted by the element's ability to form H-DNA and not by repeat number, position, or the number of mutant base pairs. DNA may therefore be a dynamic participant in the transcription of the c-myc gene.

X-ray scattering techniques

X-ray scattering techniques are a family of non-destructive analytical techniques which reveal information about the crystallographic structure, chemical composition, and physical properties of materials and thin films.

These techniques are based on observing the scattered intensity of an X-ray beam hitting a sample as a function of incident and scattered angle, polarization, and wavelength or energy.

X-ray Diffraction Techniques

X-ray diffraction yields the atomic structure of materials and is based on the elastic scattering of X-rays from the electron clouds of the individual atoms in the system. The most comprehensive description of scattering from crystals is given by the dynamical theory of diffraction.

- Single-crystal X-ray diffraction is a technique used to solve the complete structure of crystalline materials, ranging from simple inorganic solids to complex macromolecules, such as proteins.
- Powder diffraction (XRD) is a technique used to characterise the crystallographic structure, crystallite size (grain size), and preferred orientation in polycrystalline or powdered solid samples. Powder diffraction is commonly used to identify unknown substances, by comparing diffraction data against a database maintained by the International Centre for Diffraction Data. It may also be used to characterize heterogeneous solid mixtures to determine relative abundance of crystalline compounds and, when coupled with lattice refinement techniques, such as Rietveld refinement, can provide structural information on unknown materials. Powder diffraction is also a common method for determining strains in crystalline materials. An effect of the finite crystallite sizes is seen as a broadening of the peaks in an X-ray diffraction as is explained by the Scherrer Equation.
- Thin film diffraction and grazing incidence X-ray diffraction may be used to characterize the crystallographic structure and preferred orientation of substrate-anchored thin films.
- High-resolution X-ray diffraction is used to characterize thickness, crystallographic structure, and strain in thin epitaxial films. It employs parallel-beam optics.
- X-ray pole figure analysis enables one to analyze and determine the distribution of crystalline orientations within a crystalline thin-film sample.
- X-ray rocking curve analysis is used to quantify grain size and mosaic spread in crystalline materials.

Bibliography

Annie Tan: *Blueprints Notes and Cases - Biochemistry, Genetics and Embryology*, Blackwell Science, Delhi, 2004.

Balkwill, Fran: *DNA Is Here to Stay*, Carolrhoda Books, Inc., Minneapolis, 1993.

Banerjee, S : *Cell Biochemistry*, Dominant Pub, Delhi, 2009.

Bhatia, A L : *Biochemistry and Endocrinology*, Indus Valley, Delhi, 2002.

Brookes, Martin: *Get a Grip on Genetics, Time Life Books*, East Sussex, England, 1998.

Bruce, Donald : *Engineering Genesis: The Ethics of Genetic Engineering in Non-Human Animals*, London: Earthscan, 1998.

Cahill, Lisa: *Genetics, Theology, and Ethics: An Interdisciplinary Conversation*, New York: Crossroad, 2005.

Curtis, M.: *The Geometry of DNA: A Structural Revision*. London: Blue Gallery, 1996.

Diaz, E.: *Microbial Biodegradation: Genomics and Molecular Biology,* Caister Academic Press, UK, 2008.

Drlica, Karl : *Understanding DNA and Gene Cloning: A Guide for the Curious*, New York: Wiley, 1997.

Fukuyama F.: *Our Posthuman Future: Consequences of the Biotechnology Revolution,* Farrar, Straus and Giroux, NewYork, 2002.

Garza-Valdes, L.A.: *The DNA of God?* Hodder and Stoughton, London, 1998.

Hames, B. D.: *Genes and Embryos*, Oxford University Press, New York, 1989.

Jonoska, N.: *Self-Assembling DNA Graphs, DNA-Based Computers VIII*, Springer-Verlag, Berlin, 2003.

Kallenbach, N.R.: *Molecular Structure: Chemical Reactivity and Biological Activity,* Oxford University Press, Oxford, 1988.

Lewis, Ricki: *Human Genetics: Concepts and Applications,* Boston, Mass: McGraw Hill. 1994.

Mayr, E.: *Change of Genetic Environment and Evolution*, Allen and Unwin, London, 1954.

Nisha Khalsa: *Essentials of Biochemistry*, Aavishkar Pub, Delhi, 2008.

Old, R.W.: *Principles of Gene Manipulation*, London, Blackwell Scientific Publications, 1989.

Rabus, R: *Microbial Biodegradation: Genomics and Molecular Biology*, Caister Academic Press, UK, 2008.

Sturtevant, Alfred: *History of Genetics*, Harper and Row, New York, 1965.

Stwertka, Eve : *Genetic Engineering*, Franklin Watts, New York, 1982.

Stwertka, Eve and Albert: *Genetic Engineering*, Franklin Watts, New York, 1982.

Vasudev Bhatnagar: *Cell Science and Technology*, Campus Books, Delhi, 2009.

William, Kovacs D.: *An Introduction to Geotechnical Engineering*, Prentice Hall, UK, 1981.

Williams, J.G. and R.K. Patient : *Genetic Engineering*. Oxford, England: IRL Press, 1988.

Wilson, I.: *The Blood and The Shroud*. London: Weidenfeld & Nicolson, 1998.

Winston, R.: *The Future of Genetic Manipulation*, Phoenix, London, 1997.

Index

❑❑❑